Precision Irrigation for Agriculture

# Precision Irrigation for Agriculture

## Integrating Machine Learning and Optimal Control Strategies

*Bernard Twum Agyeman*

*Department of Chemical Engineering and Materials Science*
*University of Minnesota*
*Minneapolis, USA*

*Jinfeng Liu*

*Chemical and Material Engineering Department*
*University of Alberta*
*Edmonton, AB, Canada*

WILEY

This edition first published 2026


*Registered Office(s)*
John Wiley & Sons, Inc., 111 River Street, Hoboken, NJ 07030, USA
John Wiley & Sons Ltd, New Era House, 8 Oldlands Way, Bognor Regis, West Sussex, PO22 9NQ, UK

For details of our global editorial offices, customer services, and more information about Wiley products visit us at http://www.wiley.com.

The manufacturer's authorized representative according to the EU General Product Safety Regulation is Wiley-VCH GmbH, Boschstr. 12, 69469 Weinheim, Germany, e-mail: Product_Safety@wiley.com.

Wiley also publishes its books in a variety of electronic formats and by print-on-demand. Some content that appears in standard print versions of this book may not be available in other formats.



*Library of Congress Cataloging-in-Publication Data has been applied for:*

**Hardback ISBN:** 9781394288526
**ePdf ISBN:** 9781394288540
**ePub ISBN:** 9781394288533
**oBook ISBN:** 9781394288557

Cover Design: Wiley
Cover Images: Courtesy of Bernard Agyeman, © Jinfeng Liu, © debibishop/Getty Images

Set in 9.5/12.5pt STIXTwoText by Lumina Datamatics

# Contents

# List of Figures

# List of Tables

# Preface

The sustainable use of water in agriculture has emerged as an urgent concern as population growth and climate change intensify water scarcity in various regions around the world. Agriculture, which consumes close to 70% of the world's freshwater, must adopt smarter and more efficient practices to support the rising food demand while conserving freshwater resources. Precision irrigation, which can be achieved by transforming the irrigation process into a closed-loop system, offers a promising solution to this challenge by ensuring that water is delivered to crops only when and where it is required. Optimal control techniques, such as model predictive control, along with the three main machine learning paradigms—supervised, unsupervised, and reinforcement learning—have proven invaluable in devising precise irrigation strategies. However, despite significant progress in applying these techniques, existing studies have been limited to individual machine learning paradigms or have incorporated only a single machine learning approach into optimal control-based irrigation systems. This limited scope has constrained the development of a more comprehensive approach to precision irrigation. A unified framework that integrates all machine learning paradigms with optimal control offers a more effective solution. Such an approach can better address the multifaceted challenges of irrigation management, enabling more precise and resource-efficient irrigation practices.

This book aims to address a critical gap in precision irrigation by presenting a closed-loop irrigation scheduling framework that integrates optimal control with supervised, unsupervised, and reinforcement learning techniques. Specifically, the book introduces: (i) an efficient soil moisture and hydraulic parameter estimation technique that leverages remotely sensed data; (ii) a performance-triggered adaptive model reduction framework for efficient soil moisture estimation in large-scale agricultural fields; (iii) a mixed-integer model predictive control-based scheduler to comprehensively handle irrigation scheduling across diverse agricultural fields; (iv) a multi-agent learning-based model predictive control paradigm for irrigation scheduling; (v) a multi-agent reinforcement learning approach that combines centralized and decentralized agents in a semi-centralized fashion to effectively address the irrigation scheduling problem in large-scale fields; and (vi) a hierarchical integration of scheduling and control layers, enabling the use of multi-timescale information to enhance the precision of irrigation decisions.

The proposed methods draw on a wide range of tools, including optimal control, nonlinear systems, machine learning, multi-agent systems, and optimization. The effectiveness of these methods is demonstrated through multiple simulated case studies and actual field implementations. Applying these approaches to large-scale agricultural fields is expected to enhance crop yield while ensuring a sustainable use of water, labor, and energy.

Readers of this book are expected to have foundational knowledge in ordinary and partial differential equations, control theory (both linear and nonlinear), machine learning, statistics, and numerical analysis. The content is tailored for researchers, graduate students, control system engineers, and professionals in the field of irrigation. Each chapter includes discussions on real-world implementation considerations to equip practitioners and researchers with practical insights into the proposed strategies.

We would like to express our sincerest thanks to Professor Willemijn M. Appels and Dr. Mohamed Naouri of Lethbridge Polytechnic, Dr. Maik Wolleben of Skaha Labs, Professor Xunyuan Yin of Nanyang Technological University, and our friends, colleagues, and students at the University of Alberta for their invaluable input and support. In particular, we thank Professor Sirish L. Shah, Dr. Song Bo, Dr. Sarupa Debnath, Dr. Benjamin Decardi-Nelson, Dr. Soumya Sahoo, and Erfan Orouskhani for their generous contributions. We also acknowledge our colleagues at the University of Alberta for fostering a collaborative research environment. Above all, we deeply appreciate the unwavering encouragement and support of our families, without whom this endeavor would not have been possible.

July 2025

*Bernard Twum Agyeman*
*Jinfeng Liu*
*Minneapolis, Minnesota, USA and*
*Edmonton, AB, Canada*

# 1

# Introduction

## 1.1 Challenges in Irrigation and a Need for Innovation

Irrigation constitutes the largest component of freshwater use within agriculture, which itself accounts for approximately 70% of global freshwater withdrawals [1]. This substantial demand is occurring alongside growing concerns over water scarcity, a challenge driven by both climate change and population growth. Agriculture is both susceptible to these pressures and partially responsible for them, particularly due to its continued reliance on inefficient irrigation practices. In many cases, irrigation is implemented as an open-loop system, where water is applied based on fixed schedules rather than actual soil moisture conditions. The absence of feedback in such systems often results in excessive water use and significant inefficiencies.

Precision irrigation has emerged as a means of addressing the inefficiencies that characterize many conventional irrigation methods. A fundamental aspect of this approach is the transition from open-loop to closed-loop irrigation systems [2]. Closed-loop systems utilize real-time feedback, most commonly from soil moisture sensors, to guide irrigation decisions, ensuring that water is applied only when it is required. Developing such systems involves the integration of optimal control strategies alongside machine learning techniques. Optimal control provides a structured approach for determining irrigation schedules that balance agronomic objectives with resource constraints. In parallel, machine learning contributes the ability to extract insights from data and support adaptive responses to dynamic field conditions.

Although both optimal control and machine learning have shown promise in irrigation applications, further advancements are needed to meet growing demands for precision and adaptability. A key priority is the refinement of existing control and learning methods to improve the accuracy of scheduling decisions within closed-loop frameworks. Moreover, current implementations often integrate these techniques in a limited or weakly coordinated manner. For example, many systems rely on a single machine learning paradigm or embed machine learning only superficially within optimal control-based irrigation schedulers. This fragmented integration has constrained the overall effectiveness of such systems. Overcoming these limitations requires a unified framework that combines optimal control and machine learning into a cohesive, closed-loop irrigation scheduling architecture.

*Precision Irrigation for Agriculture: Integrating Machine Learning and Optimal Control Strategies*, First Edition. Bernard Twum Agyeman and Jinfeng Liu.

This book aims to advance the application of optimal control and machine learning in the development of closed-loop irrigation systems. Beyond refining these methods individually, it introduces a unified scheduling framework that integrates optimal control with all three major paradigms of machine learning: supervised, unsupervised, and reinforcement learning (RL). By embedding enhanced machine learning techniques and adapted control strategies within a single framework, the proposed approach seeks to improve water-use efficiency and support sustainable agricultural practices in the face of increasing global water constraints.

## 1.2 Current Advances and Unresolved Challenges

At its core, irrigation scheduling involves specifying the appropriate times and quantities of water application to support healthy crop development over the course of a growing season. These decisions are typically made at regular intervals, often daily or hourly, depending on the specific water requirements of the crop and the configuration of the irrigation system in use. For example, a daily scheduling approach identifies which days within a scheduling horizon require irrigation and assigns water volumes accordingly.

In settings where rainfall alone is insufficient to support plant development, effective scheduling is essential for maintaining improved crop yields. Both under- and over-irrigation present significant challenges. Inadequate water supply during key developmental stages can lead to physiological stress and diminished crop performance. On the other hand, excessive irrigation may result in surface runoff, deep percolation, or soil saturation—all of which contribute to inefficient water use and environmental degradation. These concerns highlight the importance of precision in irrigation scheduling and have motivated the development of a diverse set of tools and methodologies aimed at improving scheduling efficiency. The following section reviews major contributions, recent innovations, and emerging opportunities for advancement in this area.

In conventional approaches to irrigation scheduling, soil moisture status is commonly inferred. This inference is typically based on one of three sources: indicators of plant water stress [3–6], data from field sensor systems [7–10], or estimations of crop evapotranspiration [6, 11–13]. An irrigation event is typically triggered when the inferred soil moisture content is lower than a specified threshold. The applied water volume is then calculated with the goal of restoring the soil profile to field capacity, which is defined as the maximum moisture content the soil can retain after drainage. While these methods are appealing due to their simplicity and low computational requirements, they are inherently limited by the accuracy of the underlying estimates. Errors in soil moisture inference, whether due to model assumptions, measurement noise, or environmental variability, can lead to poorly timed or miscalculated irrigation volumes. Such inaccuracies increase the likelihood of over-irrigation, which may cause waterlogging and nutrient leaching, or under-irrigation, which can induce plant stress and reduce yields.

To improve the accuracy and reliability of irrigation schedules, recent applications have targeted process-based modeling methods [14–17]. These approaches represent interactions within the soil–crop–atmosphere system, as captured by the underlying model, to determine irrigation schedules. Mechanistic frameworks such as the soil water balance model [18], Richards' equation [16], and crop models like AQUACROP [17] are frequently employed for this purpose. While process-based models provide a more rigorous system representation compared to rule-based methods, their practical deployment is often constrained by the complexity and effort involved in calibration. In response, attention has shifted toward data-driven alternatives that infer system behavior from empirical data rather than physical principles. Black-box models, including machine learning methods such as feedforward neural networks [19–21] and recurrent neural networks [22], have been investigated as flexible tools for capturing the dynamics of crop water use and informing irrigation decisions. These models minimize the need for detailed parameterization and can generalize across field conditions when trained on high-quality data. For a comprehensive review of both conventional and model-based scheduling approaches, readers may consult [23, 24].

Approaches based on process models for irrigation scheduling can be broadly categorized into two types: those that rely on rule-based strategies and those that rely on optimization techniques. The first category, which can be referred to as nonoptimal control methods, combines process models with simple decision rules to guide irrigation [22, 23]. While these methods are relatively easy to implement and computationally inexpensive, they typically do not account for resource limitations and do not seek to optimize irrigation volumes. In contrast, optimal control-based methods formulate irrigation scheduling as an optimization problem, where decisions regarding the timing and quantity of water application are derived by maximizing or minimizing a specific performance criterion [16, 17, 25]. These approaches allow for the explicit inclusion of constraints on available water, soil moisture thresholds, and system dynamics.

Model predictive control (MPC) is a widely used optimal control technique that computes control actions by forecasting system behavior over a finite horizon, while explicitly incorporating system constraints. By anticipating future states of the Bernard Twum Agyeman system, MPC allows for irrigation plans that enhance crop performance while conserving inputs such as water and energy. A typical use case of MPC in irrigation is set-point tracking [17, 26], where the controller seeks to maintain soil moisture at a fixed target level while optimizing water use. However, recent studies have adopted a zone-control formulation [25], which seeks to determine irrigation schedules that keep soil moisture content within some optimal range. This perspective offers greater flexibility and more accurately reflects the practical objectives of irrigation scheduling—that is, to sustain soil moisture within bounds that support healthy crop development throughout the season.

Scheduling models typically incorporate discrete decisions, particularly when limited resources must be distributed across competing demands [27]. In the case of scheduling irrigation on a daily basis, deciding which specific days within a horizon should receive irrigation represents a binary choice—whether to irrigate or not on each day. As such, a full treatment of the scheduling problem within an optimal control framework (or MPC framework) must account for both discrete and continuous control variables: the former governing the timing of irrigation events, and the latter specifying the quantity of water to apply. Some studies [25, 26] address irrigation timing outside the predictive control loop using heuristic rules. While straightforward, such strategies cannot guarantee that the resulting schedule is optimal. Other implementations [17] implicitly deduce irrigation timing from the magnitude of the irrigation amount, interpreting a near-zero value as an indication to skip irrigation on a given day. Although this simplification reduces complexity, it fails to explicitly control the timing of irrigation and can misrepresent actual water needs. In agronomic cases where a small yet essential quantity of water is required, threshold-based approaches may erroneously suppress irrigation events if the required amount falls below the predefined threshold. Furthermore, empirical results indicate that these implicit strategies often produce schedules characterized by frequent low-volume irrigation events, which can drive up operational costs. While careful tuning of the MPC cost function could mitigate this issue by imposing a stronger trade-off between timing and volume, the tuning process itself is nontrivial and problem-dependent. Importantly, recent advances in computational power and optimization tools remain underutilized in this domain. There is considerable potential to incorporate discrete decision-making directly within the MPC framework to achieve more efficient and practical scheduling outcomes.

Recent progress in sensing technologies and meteorological forecasting, combined with the increased accessibility of computational resources, has created favorable conditions for the application of machine learning in irrigation scheduling. These methods are increasingly explored as alternatives to traditional rule-based systems, offering the ability to extract insights from large volumes of historical data. Machine learning provides a flexible foundation for modeling complex relationships and has been applied under three primary paradigms: supervised learning, which supports predictive modeling; unsupervised learning, which facilitates pattern discovery; and RL, which enables adaptive decision-making in uncertain environments. For an extensive overview of machine learning applications in this domain, the reader may consult [28, 29].

Methods such as $k$-means clustering [30, 31], principal component analysis [32], and fuzzy clustering [33] have been extensively applied to divide large agricultural fields into irrigation management zones (MZs). These zones are typically delineated using spatial attributes such as soil electrical conductivity, yield patterns, and topographic variation. Defined as field subregions with relatively uniform soil and crop characteristics, MZs provide a systematic means of addressing spatial variability in large-scale agricultural systems. Incorporating these zones into closed-loop irrigation schedulers enables localized control strategies, facilitating more precise water application and improved resource efficiency. However, despite the demonstrated value of unsupervised learning for zone delineation, significant challenges remain in developing zone-specific model-based irrigation controllers. A key limitation lies in the incomplete integration of essential model parameters, particularly soil hydraulic properties. Acquiring high-resolution spatial data on these parameters across large fields is often both labor-intensive and costly. One practical solution involves estimating them through inverse modeling approaches. This underscores the need for computationally efficient methods capable of inferring spatial distributions of soil hydraulic properties to support the development of MZs that are compatible with process-based irrigation models.

Soil moisture and hydraulic property estimation have been the subject of extensive research, with numerous techniques developed to infer these variables from observational data. Methods such as the extended Kalman filter (EKF) [34, 35], the ensemble Kalman filter (EnKF) [36–40], and the moving horizon estimator [41, 42] have been widely applied for this purpose. A common limitation of these approaches is their dependence on point-based soil moisture measurements, which constrains their ability to capture the spatial variability of soil hydraulic parameters across large fields. Remote sensing technologies, particularly those based on microwave sensing, offer a promising alternative by providing spatially distributed soil moisture data, enabling parameter estimation at scale. Furthermore, several studies developing estimation frameworks overlook the importance of assessing parameter estimability. Estimability, which evaluates whether specific model parameters can be uniquely determined from available measurements [43, 44], is a critical yet frequently neglected step. Its inclusion is especially important in large-scale agricultural settings, where the inability to uniquely estimate certain parameters can lead to unreliable estimates. Moreover, given the sparse and often noisy nature of soil moisture data, attempting to estimate all hydraulic parameters across the field is rarely feasible. A more practical strategy involves identifying a subset of parameters that can be estimated with sufficient accuracy. Integrating a parameter selection step into the estimation process not only enhances the robustness of the results but also reduces computational burden by lowering the dimensionality of the problem. In summary, improving current estimation methodologies requires the combined use of spatially distributed soil moisture observations, formal estimability analysis to identify reliably estimable parameters, and parameter selection techniques to focus on those that are both estimable and informative.

Reliable predictions of weather conditions and soil moisture are essential for effective irrigation scheduling, as they guide decisions about when to irrigate and how much water to apply. Supervised learning techniques have played a key role in this area by leveraging historical data to construct predictive models tailored to specific field conditions. A wide range of algorithms has been applied for this purpose, including support vector machines [45, 46], decision trees [47, 48], random forests [49, 50], *k*-nearest neighbors [51, 52], feed-forward neural networks [19–21], and long short-term memory networks [22]. Among these applications, one of the most critical challenges is accurately estimating soil moisture within the crop root zone, as this directly impacts plant water uptake and the precision of scheduling decisions. Models addressing this task often adopt ensemble learning strategies [19] or multi-input, multi-output architectures [53]. While these models are well-suited for rule-based scheduling systems, their computational complexity can hinder integration into optimization-based frameworks. This underscores the need for supervised models that are not only accurate but also computationally efficient and compatible with real-time, control-oriented irrigation scheduling.

RL has emerged as a promising tool for optimizing irrigation scheduling by enabling agents to learn effective water management strategies through interaction with their environment. Prior applications [54–57] have adopted model-free RL approaches, such as Q-learning, to develop policies that respond to changing field conditions. These approaches have demonstrated improvements in water-use efficiency and crop yield across a variety of settings. Recent developments in RL, such as hybrid actor-critic algorithms [58] and hierarchical RL architectures [59], provide mechanisms for jointly managing discrete and continuous action spaces. These methods offer a pathway to more effective control policies, where RL agents can simultaneously determine whether to irrigate and, if so, how much water to apply.

Scaling daily irrigation scheduling to large agricultural fields composed of multiple MZs introduces an added level of complexity. In this setting, scheduling must account for zone-specific decisions regarding both the timing and quantity of irrigation. One effective way to structure this problem is through a multi-agent system (MAS), in which each zone is assigned an autonomous agent responsible for making local irrigation decisions. MAS-based approaches have been explored in the irrigation domain [60], with agent-based modeling offering a common framework for simulating decentralized coordination [61, 62]. While such models support local decision autonomy, they often rely on predefined rules and lack mechanisms for adapting to dynamic environments. Recent progress in RL has demonstrated the potential to overcome this limitation by enabling agents to learn from interaction with their environment. When integrated into MAS frameworks, RL gives rise to multi-agent reinforcement learning (MARL), which supports both collaborative and independent learning among agents. This paradigm presents an opportunity for addressing the complexities of irrigation scheduling in large-scale fields that are delineated into MZs.

During the development of MARL frameworks, the initial approach in many settings is to adopt a fully observable Markov decision process formulation, where each agent is assumed to have access to the true underlying state of its environment. While this assumption simplifies the training process, it is often impractical in real-world applications such as agricultural systems. In these contexts, complete information about the system state (i.e., the spatial distribution of soil moisture in the root zone) is rarely available. Instead, agents must rely on partial observations obtained through sensor networks, which offer only limited insight into the full environmental state. As such, while it is reasonable to begin the development of MARL-based irrigation systems under the fully observable assumption, the insights gained from this simplified setup should inform the design of more realistic frameworks. These subsequent frameworks should relax the full observability assumption and incorporate mechanisms for decision-making under partial observability.

Recent advancements in sensor technology have made it feasible to obtain high-resolution weather data, including hourly forecasts. Until now, the discussion in this section has focused primarily on daily irrigation scheduling, where decisions are based on daily weather summaries and soil moisture estimates to determine both the timing and amount of irrigation. While daily data offer useful aggregated insights, they often mask important intraday fluctuations in weather variables that can significantly influence crop water requirements. To better capture these sub-daily dynamics, a more comprehensive scheduling approach would incorporate both daily and hourly weather information into the decision-making process. However, in the context of MPC, especially when the problem includes both discrete and continuous decision variables, naively combining multi-timescale data can result in a substantial increase in computational complexity. This motivates the need for computationally efficient strategies to manage and integrate information across different temporal resolutions. One promising direction comes from process systems engineering, where hierarchical control architectures are commonly used. Applied to irrigation, a similar structure could involve a scheduling layer that operates at a daily level—using daily weather forecasts to plan irrigation timing and volumes—alongside a lower-level control layer that responds to hourly weather fluctuations to ensure that the daily schedule is executed accurately. Such a hierarchical framework will allow for scheduling at longer timescales while retaining the flexibility to adapt to short-term environmental variability, thereby improving both the adaptability and efficiency of the irrigation system.

In conclusion, the current body of research on irrigation scheduling reveals several critical gaps. These include the lack of an integrated framework combining optimal control with all three machine learning paradigms, and the absence of strategies that jointly address discrete and continuous decisions in daily scheduling. Efficient, scalable methods for field-scale hydraulic parameter estimation are also underdeveloped. Moreover, supervised learning models that accurately represent soil moisture dynamics and align with control-based scheduling remain limited. RL methods capable of mixed-action control, as well as multi-agent approaches suited for zone-based field management, have yet to be fully explored. Finally, progress in hierarchical frameworks that couple scheduling with control to leverage multi-timescale information remains minimal.

## 1.3 Objectives and Organization of the Book

Motivated by the lack of a unified irrigation scheduling framework that integrates optimal control with the three machine learning paradigms, this book has the following broad objectives:

1. To design a computationally efficient approach for estimating soil hydraulic parameters at the field scale.
2. To develop an MPC-based irrigation scheduler that employs both discrete and continuous controls to address the daily irrigation scheduling problem in homogeneous fields and spatially variable fields characterized by multiple MZs.
3. To create a unified framework that integrates the MPC-based irrigation scheduler with unsupervised learning for management zone delineation (using estimated hydraulic parameters), a streamlined supervised learning model for soil moisture modeling within the delineated zones, and RL to enhance the computational efficiency of the overall scheduling framework.
4. To develop a MARL framework for daily irrigation scheduling in large-scale fields characterized by multiple MZs.
5. To develop a hierarchical framework that integrates scheduling and advanced control, enabling the use of multi-timescale information for adaptive and efficient irrigation decision-making.
6. To demonstrate the application of the unified scheduling framework in large-scale agricultural fields.

The book is organized as follows. Chapter 2 reviews fundamental concepts of soil moisture modeling in agricultural fields. This chapter introduces and derives the Richards' equation, which is used throughout this book to describe soil moisture dynamics in large-scale agricultural fields. Additionally, a numerical solution technique for the Richards' equation is presented.

In Chapter 3, machine learning and MPC are reviewed. For machine learning, the three main paradigms are introduced, followed by a detailed discussion of techniques such as feedforward neural networks, long short-term memory networks, physics-informed neural networks, $k$-means clustering, actor-critic RL methods, proximal policy optimization, and MARL. In the context of MPC, key aspects such as formulation, feasibility, and tuning are examined.

Chapter 4 presents estimation of soil moisture and hydraulic parameters in the context of large-scale agricultural fields. The chapter introduces an estimation framework that leverages soil moisture data obtained from microwave remote sensors mounted on a center pivot irrigation system. A key component of this framework is a tailored sensitivity analysis, adapted to account for the characteristics of the remotely sensed measurements, which is used to evaluate parameter estimability prior to estimation. To enhance reliability, an orthogonal projection method is applied to determine a subset of hydraulic parameters suitable for unique estimation. This is followed by a modified EKF designed to incorporate the results of the parameter selection step. The proposed approach is validated through both a simulated scenario and a real-world case study involving a segment of a large-scale field.

In Chapter 5, the computational challenges associated with using high-dimensional and complex models for soil moisture estimation in large-scale agricultural fields are addressed through a performance-triggered model reduction framework. This framework employs a structure-preserving reduction approach and dynamically adjusts the order of the reduced model based on prediction performance. This enables the system to capture evolving field dynamics while maintaining computational efficiency. The reduced-order modeling approach is integrated with a state estimation scheme based on the EKF. A simulated case study on a large-scale field is used to demonstrate the framework's ability to effectively balance estimation accuracy with computational cost.

Chapter 6 presents a mixed-integer MPC with zone control to solve the irrigation scheduling problem on a daily basis. The chapter provides tailored formulations for both uniform and spatially heterogeneous fields. To mitigate the computational demands associated with mixed-integer optimization, a data-driven modeling strategy is presented. Furthermore, the chapter introduces an alternative formulation that substitutes binary decision variables with a smooth approximation. The effectiveness of the proposed MPC approaches is illustrated through two case studies: one involving a uniform field and the other a spatially heterogeneous field.

Chapter 7 presents a framework that combines mixed-integer MPC with all three paradigms of machine learning to tackle irrigation scheduling in fields that exhibit spatial variability. The chapter begins by outlining a three-step management zone delineation strategy that employs estimated soil hydraulic parameters alongside $k$-means clustering to partition fields into zones suitable for process-based modeling. It then details the use of long short-term memory networks to construct surrogate models of soil moisture for each zone, capturing moisture variability within the root zone while accounting for changes in rooting depth over time. To solve the discrete decision component of the mixed-integer MPC, the framework incorporates decentralized hybrid proximal policy optimization agents. These agents, in tandem with a heuristic strategy, significantly reduce the computational burden of scheduling. The proposed framework is validated through an application to a section of a large-scale field and benchmarked against a standard threshold-based irrigation method.

Chapter 8 presents a semi-centralized MARL framework to address daily irrigation scheduling in fields exhibiting spatial variability. The framework adopts a hierarchical structure that separates decision-making across two levels: a centralized agent is responsible for determining the timing of irrigation events, while decentralized agents independently regulate the irrigation amounts within their respective zones. To address the non-stationarity often encountered in multi-agent settings, a state augmentation strategy is introduced. The effectiveness of the semi-centralized approach is demonstrated through its application to a spatially heterogeneous field and evaluated against the decentralized MARL method presented in Chapter 7.

Chapter 9 presents a hierarchical framework that combines semi-centralized MARL with an advanced control layer to address irrigation scheduling in large-scale, spatially variable fields. The scheduler, trained under a partially observable Markov decision process framework, constructs belief states using an EKF. Operating at the upper level of the hierarchy, the scheduler generates daily irrigation schedules by incorporating daily soil moisture estimates, weather forecasts, and crop information. The lower level consists of an MPC layer, which refines and executes these schedules using higher-frequency inputs such as hourly weather data, real-time crop conditions, and sensor feedback. The effectiveness of the proposed hierarchical approach is demonstrated through its application to a large-scale field scenario.

Finally, Chapter 10 outlines potential research directions aimed at advancing and refining the methodologies presented in the preceding chapters.

# 2

# Background on Soil Moisture Modeling

Soil moisture modeling plays an important role in the design of irrigation schedulers. This chapter aims to introduce fundamental concepts related to soil moisture modeling in agricultural fields, providing essential background for the results and concepts presented in subsequent chapters. This review is not exhaustive; rather, it is intended to equip the reader with a foundational understanding. For a more detailed discussion, readers may refer to [63, 64].

## 2.1 Introduction

An agro-hydrological system describes the movement of water between crops, soil, and the atmosphere. A key component of this system is the vadose zone, also known as the unsaturated zone, which lies between the ground surface and the water table. The water held in this zone is referred to as soil moisture, and it plays an important role in supporting plant growth. Soil moisture in the vadose zone is highly dynamic, shifting between wet conditions following rainfall or irrigation and dry states during periods of evapotranspiration, where water is extracted by plant roots and evaporates from the soil surface. Understanding how soil moisture moves vertically and horizontally in the vadose zone is essential for effectively managing water resources, particularly in agriculture, where irrigation is vital for crop yields.

In the context of simulating soil moisture dynamics in agro-hydrological systems, early approaches employed water balance models, which work on the principle of mass conservation. These models approximate inflows from precipitation and irrigation, and outflows due to evaporation and runoff. While they are still in use today and remain valuable for certain applications, water balance models simplify the process by not fully capturing the more complex dynamics of water movement through soil. They are limited in accounting for moisture variations, particularly in saturated soils where water drains quickly, and in dry soils where water becomes increasingly difficult to extract.

To accurately model the complex behavior of water in soil, physical principles such as Darcy's law and the Richards' equation are required. Darcy's law describes how water

*Precision Irrigation for Agriculture: Integrating Machine Learning and Optimal Control Strategies*,
First Edition. Bernard Twum Agyeman and Jinfeng Liu.

moves through soil based on the hydraulic gradient, while the Richards' equation integrates Darcy's law with the principle of mass conservation to describe water flow in unsaturated soils over time. These models provide a more comprehensive and accurate depiction of soil moisture dynamics by considering factors such as soil texture, porosity, and the forces acting on water within the soil, including gravity, capillary action, and pressure gradients. This makes them important tools for simulating water movement in agro-hydrological systems, where understanding moisture variation is key to efficient water management and crop growth.

## 2.2 Water Content

The state of water in soil is typically described in terms of water content, which indicates the amount of water present within the soil. Water content provides an estimate of the stored water in the soil and helps determine how much additional water is needed through irrigation to reach a desired moisture level.

Soil consists of solid particles of varying shapes and sizes, interconnected by pores. When these pores are completely filled with water, the soil is considered saturated. On the other hand, when the pores are filled with air, the soil is considered dry. Typically, soil conditions are somewhere between these two extremes, with a mix of soil particles, air, and water.

The bulk volume of soil, $V$, is the total volume of the solid particles $V_s$ and the pore spaces $V_p$, such that $V = V_s + V_p$. The **bulk density** of a soil $\rho_b$ (kg/m$^3$) is defined as the mass of solid particles, $m_s$, per bulk volume of soil:

$$\rho_b = \frac{m_s}{V} \tag{2.1}$$

Two common measures of soil moisture are **volumetric water content** and **gravimetric water content**:

1. **Volumetric Water Content** ($\theta_v$) (m$^3$ $H_2O$/m$^3$ soil): This is defined as the volume of liquid water, $V_w$, per bulk volume of soil:

$$\theta_v = \frac{V_w}{V} \tag{2.2}$$

2. **Gravimetric Water Content** ($\theta_g$) (kg $H_2O$/kg dry soil): This is defined as the mass of water, $m_w$, per mass of dry soil $m_s$:

$$\theta_g = \frac{m_w}{m_s} \tag{2.3}$$

The relationship between gravimetric water content and volumetric water content is given by the bulk density of soil and the density of water, $\rho_w$:

$$\theta_g = \frac{\rho_w \theta_v}{\rho_b} \tag{2.4}$$

While hydrologists and soil scientists typically use volumetric water content, geotechnical engineers often prefer gravimetric water content. Throughout this book, unless otherwise specified, water content will be expressed on a volumetric basis.

## 2.3 Energy Status of Soil Water

Understanding the energy status of soil water is important because it dictates the movement and availability of water within the soil, particularly in the vadose zone. Water flows from areas of high potential energy to areas of low potential energy, influencing processes such as infiltration, drainage, and plant water uptake. To effectively manage soil moisture and irrigation, it is important to quantify the total potential energy of soil water. The total soil–water potential is defined as:

$$\underbrace{\phi_T}_{\text{total soil-water potential}} = \underbrace{\phi_{SW}}_{\text{soil-water potential}} + \underbrace{\phi_G}_{\text{elevation potential}} \tag{2.5}$$

where the soil–water potential $\phi_{SW}$ accounts for the movement of water due to gradients in capillary pressure, chemical composition, temperature, and electrical potential. The primary components of the soil–water potential are defined as follows:

$$\phi_{SW} = \phi_{matric} + \phi_{pressure} + \phi_{osmotic} \tag{2.6}$$

- **Matric Potential** ($\phi_{matric}$): This component reflects the energy state of water relative to pure water at 20 °C. It arises from capillary forces, which dominate in wet soils, and adsorptive forces, which are more significant in dry soils.
- **Pressure Potential** ($\phi_{pressure}$): This accounts for soil air pressure in unsaturated regions or positive hydrostatic pressure in saturated zones.
- **Osmotic Potential** ($\phi_{osmotic}$): This component is influenced by differences in chemical concentration, which can affect the energy state of water within the soil.

The components of the total soil–water potential can be expressed in various units:

1. Potential energy per unit weight (L).
2. Potential energy per unit mass ($L^2/T^2$).
3. Potential energy per unit volume ($M/T^3$).

In computations, potential energy is commonly expressed per unit weight, where the soil–water potential is referred to as the **pressure head** ($\psi$). To convert the volume-based soil–water potential $\phi_{SW}$, to pressure head $\psi$, the following equation is used:

$$\psi = \frac{\phi_{SW}}{\rho g} \tag{2.7}$$

where $g$ is the gravitational constant and $\rho$ is the density of the fluid. When the potential energy of water in soil is expressed on a per unit weight basis, the total potential energy equation (i.e., Equation 2.5) simplifies to:

$$H = \psi + z \tag{2.8}$$

where $H$ is the total hydraulic head and $z$ is the elevation head relative to an arbitrary reference datum.

## 2.4 Soil–Water Retention Curve

The relationship between the soil moisture content ($\theta_v$) and the pressure head ($\psi$) is described by the soil–water retention curve. This curve is represented mathematically by equations that relate $\theta_v$ to $\psi$, denoted as $\theta_v(\psi)$. One of the most widely used models for the soil–water retention curve is the van Genuchten equation [65], given by:

$$\theta_v(\psi) = \theta_r + (\theta_s - \theta_r)\left[\frac{1}{1+(-\alpha\psi)^n}\right]^{1-\frac{1}{n}} \tag{2.9}$$

where $\theta_s$ is the saturated volumetric moisture content, $\theta_r$ is the residual water content, and $\alpha$ and $n$ are curve-fitting parameters that describe the soil's hydraulic properties.

The parameters $\theta_s$, $\theta_r$, $\alpha$, and $n$ vary depending on soil texture. Specialized functions, known as pedotransfer functions, can be used to relate these parameters to specific texture classes or as continuous functions of soil properties such as clay and sand content.

## 2.5 Darcy's Law

Water movement through soil is an important process in understanding soil moisture dynamics and is governed by the principles of fluid flow in porous media. One of the most fundamental equations describing fluid flow in porous media is Darcy's law. The law is named after Henry Darcy, who first formulated it based on experiments in the mid-nineteenth century. It provides a mathematical relationship that describes the flow of water through a porous medium, such as soil, in response to potential energy differences.

In essence, Darcy's law states that the flow rate (also known as flux) of water through a porous medium is directly proportional to the hydraulic gradient, which is the driving force for water movement. The hydraulic gradient consists of two components: a pressure head and a gravitational head. The constant of proportionality that relates the flux to the hydraulic gradient is known as the hydraulic conductivity, which measures the ability of soil to transmit water.

In the vertical dimension, Darcy's law can be written as:

$$Q = -K(\theta_v)\frac{\partial(\psi + z)}{\partial z} \tag{2.10}$$

where $Q$ (m/s) is the water flux, $K(\theta_v)$ (m/s) is the hydraulic conductivity, $\psi$ (m) is the pressure head, and $z$ (m) is the gravitational head, which represents the elevation above a reference point. The negative sign in Darcy's law indicates that water flows from higher to lower potential.

The hydraulic conductivity, $K(\theta_v)$, which is a function of the soil water content $\theta_v$, measures how easily water can move through the pore spaces of the soil medium. It depends on soil structure and texture; sandy soils generally have higher values of $K$, permitting water to flow through them more easily than clay soil, which have lower values of $K$. The relationship between hydraulic conductivity and soil moisture is described by empirical equations that take into account the soil's texture. A common formulation used is based on the van Genuchten model, which expresses $K(\theta_v)$ as:

$$K(\theta_v) = K_s S_e^{0.5}[1-(1-S_e^{-m})^m]^2$$
$$m = 1 - \frac{1}{n} \tag{2.11}$$
$$S_e = \frac{\theta - \theta_r}{\theta_s - \theta_r}$$

where $K_s$ is the saturated hydraulic conductivity and $S_e$ is the effective saturation. Collectively, the parameters $K_s$, $\theta_s$, $\theta_r$, $\alpha$, and $n$ are known as the van Genuchten parameters.

In Darcy's law, the total potential energy $(\psi + z)$ driving the flow of water is the sum of two components:

1. **Pressure Head** $\psi$: This represents the capillary forces in the soil, which cause water to be held in the pore spaces. The pressure head has values $\psi < 0$ for unsaturated soil and $\psi \geq 0$ for saturated soil.
2. **Gravitational Head** $z$: This represents the potential energy due to the elevation of water. The vertical depth $z$ is taken as positive in the upward direction, so that $z=0$ is the ground surface and $z<0$ is the elevation relative to the surface with greater depth into the soil.

## 2.6 Richards' Equation

The movement of water in unsaturated soils is governed by the principle of mass conservation. When water enters and leaves a given volume of soil, the rate of change in the soil's moisture content must equal the difference between the inflow and outflow. This principle is captured mathematically by the Richards' equation, which combines Darcy's law with the concept of conservation of mass to describe water flow in unsaturated porous media, such as soil.

Consider a small control volume of soil with horizontal area $\Delta x \Delta y$ and thickness $\Delta z$, as shown in Figure 2.1. In this control volume, water is assumed to flow only in the vertical dimension. The mass flux of water (kg/s) entering the soil across the cross-sectional area $\Delta x \Delta y$ is:

$$\rho_w Q_{\text{in}} \Delta x \Delta y \tag{2.12}$$

where $\rho_w$ is the density of water.

Similarly, the mass flux leaving the soil is:

$$\rho_w Q_{\text{out}} \Delta x \Delta y \tag{2.13}$$

The change in water stored within the control volume is proportional to the rate of change of the volumetric moisture content $\theta_v$, such that the mass of water in the soil is:

$$\rho_w \theta_v \Delta x \Delta y \Delta z \tag{2.14}$$

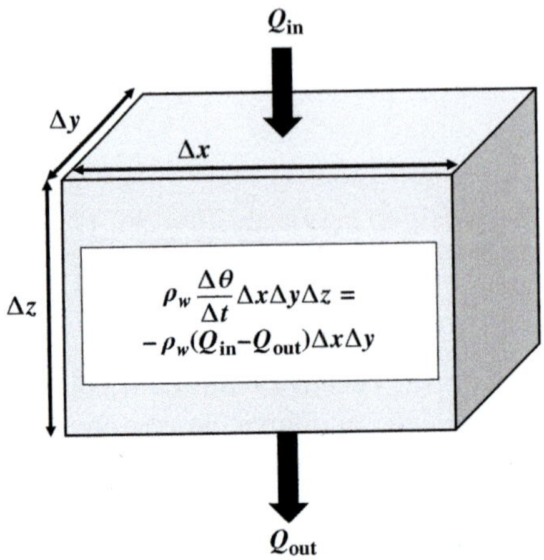


**Figure 2.1** Soil water balance for a soil control volume with fluxes $Q_{in}$ entering and $Q_{out}$ leaving the control volume.

Therefore, the conservation of mass can be written as:

$$\rho_w \frac{\Delta\theta_v}{\Delta t}\Delta x\Delta y\Delta z = -\rho_w(Q_{in} - Q_{out})\Delta x\Delta y \tag{2.15}$$

This simplifies to:

$$\frac{\Delta\theta_v}{\Delta t} = -\frac{\Delta Q}{\Delta z} \tag{2.16}$$

In differential form, the equation becomes:

$$\frac{\partial\theta_v}{\partial t} = -\frac{\partial Q}{\partial z} \tag{2.17}$$

By substituting Darcy's law for $Q$, the Richards' equation is obtained as:

$$\frac{\partial\theta_v}{\partial t} = \frac{\partial}{\partial z}\left[K(\theta_v)\frac{\partial(\psi + z)}{\partial z}\right] \tag{2.18}$$

Equation (2.18) is known as the mixed form of the Richards' equation, as it involves both the time rate of change of moisture content $\theta_v$ and the gradient of the pressure head $\psi$.

To solve the Richards' equation more easily under certain conditions, it can be rewritten in terms of the pressure head, which becomes the dependent variable. This transformation is done by expressing the time rate of change of the soil moisture content $\theta_v$ in terms of the pressure head $\psi$ using the specific moisture capacity $C(\psi)$ (1/m), which is defined as:

$$C(\psi) = \frac{\partial\theta_v}{\partial\psi} \tag{2.19}$$

Thus, the left-hand side of Equation (2.18) becomes:

$$\frac{\partial\theta}{\partial t} = C(\psi)\frac{\partial\psi}{\partial t} \tag{2.20}$$

Equation (2.18) is now rewritten as:

$$C(\psi)\frac{\partial\psi}{\partial t} = \frac{\partial}{\partial z}\left[K(\psi)\frac{\partial(\psi+z)}{\partial z}\right] \tag{2.21}$$

Equation (2.21) is the head-based form of the Richards' equation, which is applicable to both saturated and unsaturated soil conditions and is widely used in modeling soil water movement.

The van Genuchten model provides empirical relationships of $C(\psi)$ and $K(\psi)$, which are required to solve Equation (2.21). The hydraulic conductivity $K(\psi)$ is expressed as:

$$K(\psi) = K_s\left[(1+(-\alpha\psi)^n)^{-\left(\frac{n-1}{n}\right)}\right]^{\frac{1}{2}} \times \left[1-\left[1-\left[(1+(-\alpha\psi)^n)^{-\left(\frac{n-1}{n}\right)}\right]^{\frac{n}{n-1}}\right]^{\frac{n-1}{n}}\right]^2, \text{ for } \psi < 0$$

$$K_s, \text{ for } \psi >= 0 \tag{2.22}$$

And the specific moisture capacity $C(\psi)$ is given by:

$$C(\psi) = (\theta_s - \theta_r)\,\alpha n\left[1-\frac{1}{n}\right](-\alpha\psi)^{n-1}[1+(-\alpha\psi)^n]^{-\left(2-\frac{1}{n}\right)}, \text{ for } \psi < 0\; S_s, \text{ for } \psi >= 0 \tag{2.23}$$

where $S_s$ (1/m) is the specific storativity coefficient of the porous medium under positive pressure.

## 2.6.1 Solution of the Richards' Equation

Due to the highly nonlinear nature of the terms $K(\psi)$ and $C(\psi)$, obtaining an analytical solution to the Richards' equation is highly complex. As a result, numerical methods are typically employed to solve the equation. In these numerical methods, finite difference approximations are used to discretize the partial differential equation, and the nonlinear terms $K(\psi)$ and $C(\psi)$ are often linearized to simplify the solution process.

There are various numerical techniques for solving the Richards' equation, each with its specific strengths. A commonly used method applies a central difference scheme to approximate the spatial derivatives, coupled with an implicit scheme for the temporal derivatives. This numerical approach will be discussed in detail in the next section.

### 2.6.1.1 Finite Difference Approximation

In this numerical method, the central difference scheme is applied to approximate the spatial derivatives in the Richards' equation. This approach ensures second-order accuracy for the spatial terms by calculating the difference between values at adjacent nodes. Additionally, an implicit scheme, specifically the backward differentiation formula, is employed to approximate the temporal derivative. The implicit scheme offers stability, especially for stiff equations like the Richards' equation.

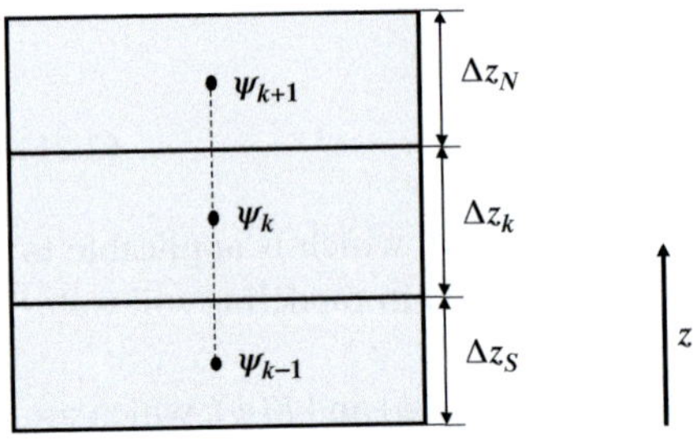


**Figure 2.2** A schematic diagram of the spatial discretization.

After performing the discretizations in both the temporal and spatial domains, the final representation of the Richards' equation is written as:

$$C_k^{p+1}(\psi)\left(\frac{\psi_k^{p+1}-\psi_k^p}{\Delta t}\right)=\left(\frac{1}{\Delta z_k}\left[K^p_{k+\frac{1}{2}}(\psi)\left(\frac{\psi_{k+1}^{p+1}-\psi_k^{p+1}}{\Delta z_N}+1\right)\right.\right.$$
$$\left.\left.-K^p_{k-\frac{1}{2}}(\psi)\left(\frac{\psi_k^{p+1}-\psi_{k-1}^{p+1}}{\Delta z_S}+1\right)\right]\right) \tag{2.24}$$

where $k\in[1,N_z]$, $N_z$ is the number of nodes in the $z$-direction, $p$ represents the time level, and $\Delta t$ is the time step size. A schematic diagram of the spatial discretization is shown in Figure 2.2. The terms $\Delta z_k$, $\Delta z_N$, $\Delta z_S$, $K^p_{k+\frac{1}{2}}$, and $K^p_{k-\frac{1}{2}}$ are defined as follows:

$$\Delta z_k=\frac{1}{2}(z_{k+1}-z_{k-1}) \tag{2.25}$$

$$\Delta z_N=z_{k+1}-z_k \tag{2.26}$$

$$\Delta z_S=z_k-z_{k-1} \tag{2.27}$$

$$K^p_{k+\frac{1}{2}}(\psi)=0.5\,(K^p(\psi_k)+K^p(\psi_{k+1})) \tag{2.28}$$

$$K^p_{k-\frac{1}{2}}(\psi)=0.5\,(K^p(\psi_k)+K^p(\psi_{k-1})) \tag{2.29}$$

For the numerical solution to be physically meaningful, appropriate boundary conditions must be applied. These conditions describe how the soil water behaves at the boundaries of the simulation domain, both at the surface (upper boundary) and at depth (lower boundary). The following are common boundary conditions used in the Richards' equation:

1. **Dirichlet Boundary Condition**: A Dirichlet boundary condition is applied when the pressure head at the boundary is prescribed. For instance, at the surface during ponding, the soil becomes saturated, and the pressure head is fixed to a known value. This boundary condition is defined as:

   $$\psi(t)=\psi_{\text{boundary}}(t)$$

   where $\psi_{\text{boundary}}(t)$ is the pressure head at the boundary (top/bottom).
2. **Neumann Boundary Condition**: The Neumann boundary condition is applied when the water flux at the boundary is prescribed. This boundary condition is useful for modeling processes like infiltration due to processes such as rainfall or irrigation at the

surface of the field. The Neumann boundary condition can also be used to model drainage at the lower boundary. Mathematically, it can be expressed as:

$$K(\psi)\frac{\partial(\psi + z)}{\partial z} = q_{\text{boundary}}(t)$$

where $q_{\text{boundary}}(t)$ is the prescribed flux at the boundary.

At the lower boundary, a special case of the Neumann boundary condition that is typically applied during the numerical solution of the Richards' equation is the **free drainage boundary condition**. The free drainage boundary condition assumes that water drains freely, resulting in zero pressure gradient at the boundary. Mathematically, this condition is expressed as:

$$\frac{\partial \psi}{\partial z} = 0$$

For a particular time level $p+1$, applying Equation (2.24) together with the boundary conditions to all the spatial nodes in the soil column results in a system of nonlinear equations. These equations can be solved iteratively using numerical methods such as Newton's method to find the pressure head at each spatial node for time level $p+1$. Once the pressure head is obtained, the corresponding volumetric soil moisture at time level $p+1$ is computed using Equation (2.9). This process is repeated for each time step until the end of the simulation period.

### 2.6.2 Sink Term

During the numerical simulation of the Richards' equation, an additional term, referred to as the sink term, is subtracted to account for other water fluxes. In the context of agro-hydrological systems, the sink term, denoted as $S(\psi, z)$, represents the root water extraction rate, which describes the volume of water taken up by plant roots per unit bulk volume of soil per unit time.

One widely used empirical model to represent the root water extraction rate is the Feddes model [66]. In this model, the root water extraction rate is calculated at various soil depths using the potential transpiration rate, the rooting depth, and the local prevailing soil water pressure head. The Feddes model is mathematically expressed as:

$$S(\psi, z) = \alpha(\psi) S_{\max}(\psi, z) \tag{2.30}$$

where $S_{\max}(\psi, z)$ ($\text{m}^3/\text{m}^3/\text{s}$) is the maximum possible root extraction rate when soil water is not limiting and $\alpha(\cdot)$ is a dimensionless water stress reduction factor.

The water stress reduction factor $\alpha(\psi)$ accounts for the reduction in root water uptake due to inadequate soil moisture, and it is calculated as follows:

$$\alpha(\psi) = \begin{cases} 0 & \psi \leq \psi_1 \\ \dfrac{\psi - \psi_1}{\psi_2 - \psi_1} & \psi_1 \leq \psi \leq \psi_2 \\ 1 & \psi_2 \leq \psi \leq \psi_3 \\ \dfrac{\psi_w - \psi}{\psi_w - \psi_3} & \psi_3 \leq \psi \leq \psi_w \end{cases} \tag{2.31}$$

where $\psi_1$ is the pressure head below which roots start to extract water from the soil, $\psi_2$ and $\psi_3$ represent the optimal range for water uptake, and $\psi_w$ is the permanent wilting point pressure head [67].

The Feddes model also defines two approaches for determining $S_{\max}(\psi, z)$: one that assumes no variation with depth and one that incorporates depth dependence. For example, in the depth-independent case, $S_{\max}(\psi, z)$ can be expressed as:

$$S_{\max}(\psi, z) = \frac{\mathrm{TP}}{L} \tag{2.32}$$

where TP (m/s) is the potential transpiration rate and $L$ (m) is the rooting depth.

The potential transpiration rate TP (m/s), which is defined as the maximum rate at which plants can transpire when soil moisture is not limiting, is computed by:

$$\mathrm{TP} = \mathrm{ETP} - \mathrm{EV} \tag{2.33}$$

where ETP (m/s) is potential evapotranspiration and EV (m/s) is the potential evaporation rate.

The potential evaporation rate EV, which quantifies the amount of water evaporating from the soil surface, is calculated following the method outlined by [68]:

$$\mathrm{EV} = \mathrm{ETP} \times \exp(-0.623\mathrm{LAI}) \tag{2.34}$$

where LAI ($m^2/m^2$) is the leaf area index representing the total leaf area of plants per unit ground area. The LAI indicates the extent of ground coverage by plant leaves, which directly influences the rate of water evaporation from the soil.

ETP (m/s) represents the combined effect of potential evaporation and transpiration from a vegetated surface under ideal conditions, is calculated as:

$$\mathrm{ETP} = K_c \times \mathrm{PET} \tag{2.35}$$

where PET (m/s) is the reference evapotranspiration and $K_c$ is the crop coefficient. The PET is a standard measure that estimates the ETP from a reference crop (typically grass), and it is calculated using meteorological data and the Penman–Monteith equation. The $K_c$ is a factor used to adjust the PET to account for the specific water needs of different crops. This coefficient varies based on crop type, growth stages, and local conditions, and is typically derived from empirical relationships and daily weather data.

## 2.7 The General Form of the Richards' Equation

While the derivation of the Richards' equation in Section 2.6 focused on vertical water fluxes for simplicity, there are scenarios where lateral fluxes become significant and must be accounted for when solving the equation. In many cases, such as in heterogeneous or sloping fields, lateral water movement may dominate the overall flux, making it essential to incorporate these horizontal components into the Richards' equation. Additionally, while the Cartesian coordinate system was used in the initial derivation, other coordinate systems, such as cylindrical coordinates, are more appropriate for specific applications. For instance, in agro-hydrological systems equipped with center pivot irrigation systems, the cylindrical coordinate system better reflects the circular nature of the irrigated area.

The above considerations motivate the need to express the Richards' equation in a more general form, one that accounts for both vertical and lateral water fluxes without assuming a specific coordinate system. This generalization allows the equation to be applied to a wide range of hydrological scenarios. The Richards' equation in its generalized form is expressed as:

$$\frac{\partial \theta_v}{\partial t} = C(\psi)\frac{\partial \psi}{\partial t} = \nabla \cdot (K(\psi)\nabla(\psi + z)) - S(\psi, z) \tag{2.36}$$

This general form can then be tailored to specific soil domains by expressing the gradient ($\nabla$) and divergence ($\nabla\cdot$) operators in the appropriate coordinate system. Furthermore, fluxes in less significant directions can be neglected, focusing on the dominant water fluxes.

For example, for a field that is equipped with a center pivot irrigation system, which results in a cylindrical soil control volume, it is more suitable to express the Richards' equation in cylindrical coordinates. To express the Richards' equation in cylindrical coordinates, the following steps can be employed:

1. **Expressing the Gradient ($\nabla$) and Divergence ($\nabla\cdot$) Operators in Cylindrical Coordinates**: In cylindrical coordinates $(r, \theta, z)$, where $r$ is the radial direction, $\theta$ is the azimuthal angle, and $z$ is the axial direction, the gradient operator is expressed as:

$$\nabla := \frac{\partial}{\partial r}\hat{r} + \frac{1}{r}\frac{\partial}{\partial \theta}\hat{\theta} + \frac{\partial}{\partial z}\hat{z} \tag{2.37}$$

The divergence of a continuously differentiable vector field $F$ in cylindrical coordinates is given by:

$$\nabla \cdot F := \frac{1}{r}\frac{\partial}{\partial r}(rF_r) + \frac{1}{r}\frac{\partial F_\theta}{\partial \theta} + \frac{\partial F_z}{\partial z} \tag{2.38}$$

where $F_r$, $F_\theta$, and $F_z$ are the components of the vector field $F$ in the radial, azimuthal and axial directions, respectively.

2. **Applying the Operators to the Generalized Richards' Equation**: By substituting these expressions for the gradient and divergence operators into Equation (2.36), the Richards' equation in cylindrical coordinates becomes:

$$C(\psi)\frac{\partial \psi}{\partial t} = \frac{1}{r}\frac{\partial}{\partial r}\left[rK(\psi)\frac{\partial \psi}{\partial r}\right] + \frac{1}{r}\frac{\partial}{\partial \theta}\left[\frac{K(\psi)}{r}\frac{\partial \psi}{\partial \theta}\right] + \frac{\partial}{\partial z}\left[K(\psi)\left(\frac{\partial \psi}{\partial z} + 1\right)\right] - S(\psi, z) \tag{2.39}$$

## 2.8 Conclusion

In this chapter, fundamental concepts of soil moisture modeling were briefly reviewed. The chapter began with an overview of methods for quantifying soil moisture, followed by a discussion on the energy status of water in soil. Darcy's law was introduced and subsequently used to derive the Richards' equation. A description of the sink term in the Richards' equation was then presented, followed by an outline of a numerical approach to solve the equation. The chapter concluded with a discussion of the general form of the Richards' equation.

# 3

# Background on Machine Learning and Model Predictive Control

Machine learning and model predictive control (MPC) have emerged as essential components in the development of closed-loop irrigation scheduling systems. This chapter introduces the core principles of both MPC and machine learning, offering a foundation for understanding the techniques applied in subsequent chapters. This discussion is not exhaustive; instead, it aims to offer essential insights for the content in the subsequent chapters of the book. For a comprehensive treatment of MPC, readers may refer to [69]. For an in-depth discussion of machine learning, particularly supervised and unsupervised learning techniques, readers may refer to [70]. Additionally, for a detailed study of reinforcement learning (RL), readers may consult [71].

## 3.1 Machine Learning

Machine Learning (ML) is a branch of artificial intelligence focused on designing algorithms that enable computers to learn from data and improve their performance on tasks over time. Unlike traditional programming, where computers operate according to fixed instructions, ML systems autonomously identify patterns, make predictions, and enhance their accuracy through exposure to data. In irrigation scheduling, for instance, ML can be instrumental in developing predictive models for soil moisture content. By training on historical data—such as temperature, humidity, rainfall, soil type, and crop characteristics—an ML model can learn to recognize relationships between these factors and soil moisture levels. Over time, the model becomes capable of forecasting future moisture levels, allowing it to recommend irrigation when soil moisture drops below a predefined threshold.

ML encompasses three primary paradigms: supervised learning, unsupervised learning, and RL. Each paradigm provides distinct methods for learning from data. The following sections will examine these paradigms in detail and discuss selected techniques within each paradigm.

### 3.1.1 Supervised Machine Learning

In supervised learning, a model is trained on labeled data, where each input is paired with the corresponding desired output. The model learns to predict outputs for new, unseen data based on these labeled examples. This process occurs under "supervision," akin to a teacher

*Precision Irrigation for Agriculture: Integrating Machine Learning and Optimal Control Strategies*,
First Edition. Bernard Twum Agyeman and Jinfeng Liu.

guiding the model's learning. Supervised learning algorithms are generally divided into two main types:

1. **Regression:** Regression algorithms are used to predict continuous values by learning a function that maps input features to a numeric output. Common regression algorithms include linear regression, polynomial regression, support vector regression, decision tree regression, and random forest regression.
2. **Classification:** Classification algorithms predict categorical outcomes by mapping input features to a probability distribution over discrete classes. Widely used classification algorithms include logistic regression, support vector machines, decision trees, and random forests.

#### 3.1.1.1 Feedforward Neural Networks

A feedforward neural network is among the simplest forms of artificial neural networks, where information flows in a single direction—from the input layer, through any hidden layers, and finally to the output layer. The architecture of a feedforward neural network, as illustrated in Figure 3.1, comprises three main types of layers:

- **Input Layer:** This initial layer consists of neurons that receive the raw input data. The number of neurons in the input layer corresponds to the dimensionality of the input data.
- **Hidden Layer(s):** Positioned between the input and output layers, hidden layers are the computational core of the network. Each neuron in a hidden layer receives weighted inputs from the previous layer, applies an *activation function*, and passes the result to the next layer. Activation functions introduce nonlinearity into neural networks, allowing them to model complex patterns in data. Common activation functions include the sigmoid, tanh (hyperbolic tangent), and rectified linear unit (ReLU). A feedforward network can have multiple hidden layers, allowing for deeper and more complex representations.
- **Output Layer:** This final layer produces the network's prediction based on the processed inputs. The number of neurons in the output layer depends on the specific task—for example, one neuron for a binary classification or multiple neurons for a multi-class problem.

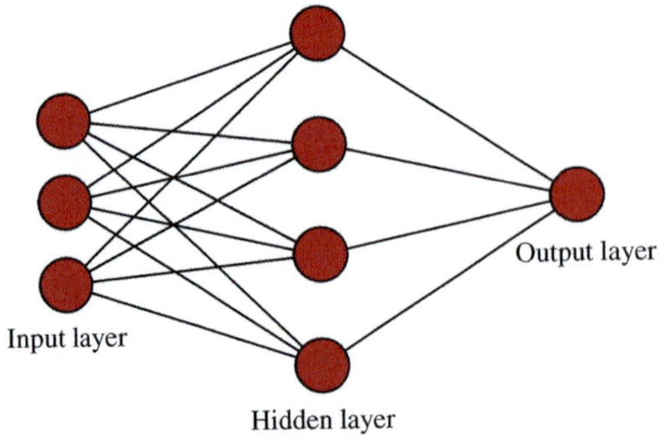


**Figure 3.1** A schematic of a feedforward neural network.

Each neuron in a given layer connects to every neuron in the subsequent layer, forming a fully connected network. The strength of these connections is determined by weights, which are adjusted through a learning process.

##### 3.1.1.1.1 Operation of a Feedforward Neural Network

The operation of a feedforward neural network consists of two main phases: the feedforward phase and the backpropagation phase.

- **Feedforward Phase:** In this phase, input data flows sequentially through the network, from the input layer through each hidden layer, until reaching the output layer. At each hidden layer, the network calculates the weighted sum of inputs and applies an activation function to introduce nonlinearity. This process continues until a prediction is generated at the output layer.
- **Backpropagation Phase:** Following a prediction, the network computes the error, defined as the difference between the predicted output and the actual output. This error is propagated backward through the network to update the weights, aiming to minimize the prediction error. Weight adjustments are typically made using the *gradient descent optimization algorithm.*

##### 3.1.1.1.2 Feedforward Neural Network Training

Training a feedforward neural network involves using a dataset to iteratively adjust the weights between neurons. By repeatedly passing over the dataset—often through multiple epochs—the weights are optimized to minimize prediction errors, a process known as gradient descent. This iterative training continues until the network achieves satisfactory performance on the data.

#### 3.1.1.2 Long Short-term Memory Networks

Long short-term memory (LSTM) networks [72] are a special type of recurrent neural network (RNN) designed specifically for sequential data processing. Traditional RNNs maintain a single hidden state across time steps, which often limits their ability to capture long-term dependencies. LSTMs address this challenge by introducing a cell state, a dedicated memory that retains information over extended periods. This capability makes LSTM architectures well-suited for complex tasks like language translation, speech recognition, and time series forecasting.

The LSTM architecture, as illustrated in Figure 3.2, features a cell state which is controlled by three types of gates—the input gate, forget gate, and output gate. These gates manage the flow of information through the cell state:

- **Input Gate:** Determines what new information should be added to the cell state.
- **Forget Gate:** Controls which information is removed from the cell state.
- **Output Gate:** Regulates the information to be output from the cell state.

Each gate is made up of a sigmoid activation function $\sigma$ and a pointwise multiplication operation $\otimes$. These gates enable LSTM networks to selectively retain or discard information. Additionally, LSTMs maintain a hidden state, which serves as the short-term memory. This hidden state is updated at each time step, based on the current input, previous hidden state, and the cell state's current value.

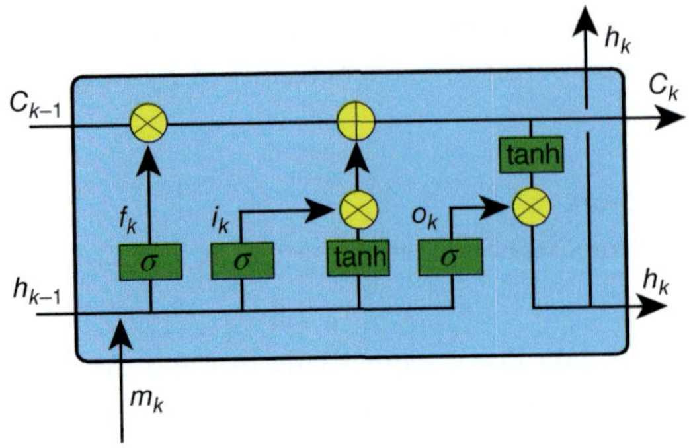


**Figure 3.2** A schematic of a long short-term memory network where $m_k$, $C_k$, and $h_k$ represent the input sequence, the cell state, and the hidden state at time step $k$, respectively.

Given an input sequence $m_k$, where $k = 1, 2, ..., T$ with $T$ representing the length of the input sequence, the LSTM processes the inputs $m_k$ to predict an output sequence $y_k$. At each time step, the network iterates through the following equations:

$$i_k = \sigma(w_i m_k + U_i h_{k-1} + b_i) \tag{3.1}$$

$$f_k = \sigma(w_f m_k + U_f h_{k-1} + b_f) \tag{3.2}$$

$$o_k = \sigma(w_o m_k + U_o h_{k-1} + b_o) \tag{3.3}$$

$$\tilde{C}_k = \tanh(w_c m_k + U_c h_{k-1} + b_c) \tag{3.4}$$

$$C_k = f_k \otimes C_{k-1} + i_k \otimes \tilde{C}_k \tag{3.5}$$

$$h_k = o_k \otimes \tanh(C_k) \tag{3.6}$$

$$y_k = w_y h_k + b_y \tag{3.7}$$

The terms $w_o$, $w_f$, and $w_i$ refer to the weights for the output, forget, and input gates, respectively. The variables $C_k$ and $h_k$ represent the cell state and hidden state at time step $k$. The matrices $U_i$, $U_f$, and $U_o$ are the weights connecting the input, forget, and output gates to the hidden state. The bias vectors associated with each gate are denoted by $b_i$, $b_f$, and $b_o$. The LSTM's output is computed using Equation (3.7), where $w_y$ and $b_y$ are the weight matrix and bias vector, respectively.

#### 3.1.1.3 Physics-informed Neural Networks

Physics-informed neural networks (PINNs) represent an advanced approach to neural network regularization by embedding physical laws directly into the learning process. PINNs leverage partial differential equations or other physical constraints to guide the network's learning, enabling it to generalize well even in instances where data is limited or noisy. This approach is particularly useful for modeling real-world physical systems, where observational data may be sparse or difficult to obtain.

Traditional neural networks rely heavily on abundant labeled data to capture underlying patterns accurately, which can lead to overfitting if the data is limited. PINNs address this limitation by encoding prior physical knowledge in the form of governing equations, thereby constraining the model to follow known behaviors of the system.

The distinctive feature of PINNs is their composite loss function, which combines data loss and physics loss terms. This composite loss function allows the model to balance fidelity to observed data with adherence to the underlying physical laws of the system.

#### 3.1.1.3.1 Data Loss Term

The data loss term, $L_{\text{data}}$, measures the discrepancy between the model's predictions and the observed data points. Given a dataset $\{x_j, y_j\}_{j=1}^{N}$, where $y_j$ represents the observed value corresponding to input $x_j$, and $f(x;\theta)$ denotes the neural network's prediction, the data loss is typically defined as:

$$L_{\text{data}} = \frac{1}{N}\sum_{j=1}^{N}\left(y_j - f(x_j;\theta)\right)^2 \tag{3.8}$$

where $\theta$ represents the network's parameters. This loss term encourages the model to closely fit the observed data.

#### 3.1.1.3.2 Physics Loss Term

The physics loss term, $L_{\text{physics}}$, enforces the network's alignment with the physical laws governing the system. Suppose the system behavior is described by a differential equation $g(x,y) = 0$, where $y$ is approximated by $f(x;\theta)$. To incorporate this physical law, PINNs introduce a set of collocation points, $\{x_i\}_{i=1}^{M}$, at which the differential equation residuals are evaluated. The physics loss is then defined as:

$$L_{\text{physics}} = \frac{1}{M}\sum_{i=1}^{M}\left(g(x_i, f(x_i;\theta))\right)^2 \tag{3.9}$$

This term penalizes deviations from the prescribed physical law, ensuring that the model's predictions respect known system behaviors.

#### 3.1.1.3.3 Composite Loss Function

The total loss function for a PINN, denoted $L_{\text{total}}$, combines the data and physics loss terms, with a weighting parameter $\lambda$ controlling the emphasis on physical consistency:

$$L_{\text{total}} = L_{\text{data}} + \lambda L_{\text{physics}} \tag{3.10}$$

where $\lambda$ adjusts the relative importance of the physics loss. A larger $\lambda$ enforces stronger adherence to the physical law, while a smaller $\lambda$ emphasizes closer alignment with the observational data.

### 3.1.2 Unsupervised Machine Learning

Unsupervised learning is a branch of machine learning that focuses on analyzing unlabeled data. Unlike supervised learning, where each data point is labeled with a specific category or outcome, unsupervised learning algorithms independently uncover patterns and relationships within the data. This type of machine learning operates without human supervision, enabling models to autonomously derive insights and uncover underlying structures in the data.

Generally, three main types of algorithms are used for unsupervised learning tasks:

1. **Clustering:** Clustering is the process of grouping unlabeled data into clusters based on inherent similarities. The objective is to identify patterns and relationships within the data, allowing the model to organize data into meaningful groups without prior knowledge of the data's content. Common clustering algorithms include $K$-means clustering, hierarchical clustering, and density-based clustering.
2. **Dimensionality Reduction:** Dimensionality reduction is the process of reducing the number of features in a dataset while preserving as much meaningful information as possible. This technique is valuable for improving the performance of machine learning algorithms, especially when handling large datasets with many features. Common dimensionality reduction methods include principal component analysis, linear discriminant analysis, and nonnegative matrix factorization.
3. **Association Rule Learning:** Association rule learning, also known as association rule mining, is a widely used technique in unsupervised machine learning for uncovering relationships or associations between variables within large datasets. This rule-based approach identifies patterns that reveal valuable correlations among variables, with applications in areas such as recommendation systems. Common methods for association rule learning include the Apriori and Eclat algorithms.

#### 3.1.2.1 *K*-means Clustering

$K$-means clustering is an unsupervised learning algorithm widely used for data clustering tasks. The algorithm aims to identify $K$ distinct clusters within the data, where $K$ represents the number of clusters specified by the user. As one of the most popular clustering methods, $K$-means is known for its efficiency, simplicity, and effectiveness.

This iterative, centroid-based algorithm partitions the dataset into clusters based on the distance between data points and their respective centroids. Each centroid, or cluster center, represents the mean of all data points within its cluster, providing a central reference point for grouping similar data points together. In this way, $K$-means clustering ensures that each data point belongs exclusively to one cluster. The algorithm can be summarized in the following steps:

1. **Initialization:**
   - Select the number of clusters, $K$.
   - Randomly initialize $K$ centroids, each representing the center of a cluster.
2. **Cluster Assignment (Expectation Step):**
   - For each data point in the dataset, calculate the distance between the data point and each of the $K$ centroids.
   - Assign each data point to the nearest centroid, forming $K$ clusters.
3. **Centroid Update (Maximization Step):**
   - Recalculate the centroid of each cluster by taking the mean of all data points assigned to that cluster.
4. **Convergence Check:**
   - Repeat steps 2 and 3 until the centroids no longer change significantly (i.e., they converge) or until a predefined number of iterations is reached.

Determining the optimal number of clusters, $K$, is crucial in $K$-means clustering as it directly impacts clustering quality. A common approach is the *Elbow Method*, a graphical technique that identifies the optimal number of clusters by plotting the *within-cluster sum of squares*. The optimal $K$ is typically chosen at the "elbow" point, where the within-cluster sum of squares begins to level off.

### 3.1.3 Reinforcement Learning

RL is a machine learning paradigm in which an agent learns how to map situations to actions with the goal of maximizing a cumulative numerical reward. Unlike traditional learning methods, the agent is not directly instructed on which actions to take; rather, it must explore and evaluate different actions to discover those that yield the highest reward. One unique aspect of RL is that an action may influence not only the immediate reward but also subsequent situations and, consequently, all future rewards. This approach is characterized by two key features: trial-and-error search and the concept of delayed reward, both of which are foundational to the RL framework.

RL can be formalized as an optimal control problem for an incompletely known Markov decision process. This formalism captures the key components of a learning agent's interaction with its environment over time, emphasizing three essential aspects: the agent's perception of the environment's state, the actions it takes to influence this state, and the pursuit of a defined goal or goals associated with the environment's state.

#### 3.1.3.1 Finite Markov Decision Processes

Markov decision processes (MDPs) provide a foundational framework for understanding how an agent learns to achieve a goal through interaction. In this setup, the agent, which is the learner and decision-maker, interacts continuously with its environment—everything outside the agent that it can influence through its actions. This interaction is cyclical: the agent selects actions, the environment responds with new situations, and rewards are generated (see Figure 3.3). Rewards are numerical values indicating how favorable certain actions are in achieving the agent's goal, and the agent aims to maximize these rewards over time through strategic action selection.

The agent and environment interact in a sequential manner at discrete time steps, represented by, $t = 0, 1, 2, 3, \ldots$. At each time step $t$, the agent receives a representation of the environment's state, $S_t$, and selects an action, $A_t$ based on this information. In the next time

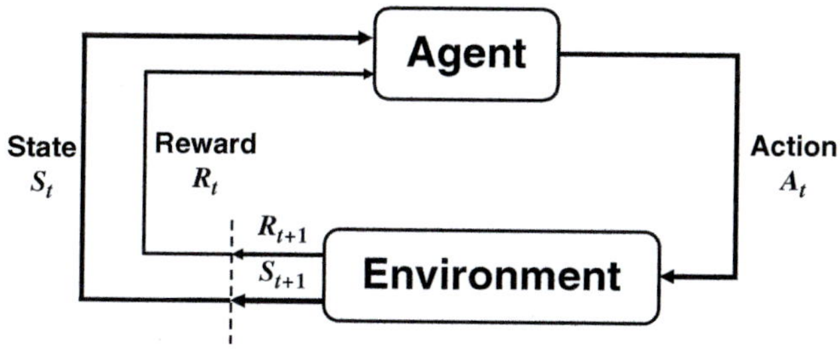


**Figure 3.3** A schematic of the agent–environment interaction in a Markov decision process.

step, partly as a result of its action, the agent receives a numerical reward, $R_{t+1}$, and transitions to a new state, $S_{t+1}$. This sequence, or trajectory, of interaction can be represented as follows:

$$S_0, A_0, R_1, S_1, A_1, R_2, \ldots \tag{3.11}$$

In a finite MDP, the sets of states $S$, actions $A$, and rewards $R$ are finite. The state $S_t$ and rewards $R_t$, considered as random variables, follow defined discrete probability distributions dependent only on the previous state and action. For given values $s' \in S$ and $r \in R$, the probability of transitioning to state $s'$ with reward $r$ at time $t$ given the prior state $s$ and action $a$, is expressed as:

$$p(s', r|s, a) := Pr\{S_t = s', R_t = r|S_{t-1} = s, A_{t-1} = a\} \tag{3.12}$$

The function $p$ fully defines the environment's dynamics within the MDP framework, capturing how the environment evolves in response to the agent's actions.

#### 3.1.3.2 Rewards, Returns, and Episodes

In RL, an agent's purpose or objective is defined through a reward signal, a numerical value, $R_t \in R$ transmitted from the environment to the agent at each time step. This reward signal reflects immediate feedback from the environment; however, the agent's true goal is to maximize the long-term accumulation of these rewards, known as the cumulative reward.

To achieve this, RL seeks to maximize the expected return, denoted as $G_t$, which represents a cumulative function of the reward sequence. In its simplest form, the return is the sum of rewards over a sequence, defined as:

$$G_t := R_{t+1} + R_{t+2} + R_{t+3} + \cdots + R_T \tag{3.13}$$

where $T$ is the final time step. This approach is well-suited to scenarios with a clear endpoint, where the interaction between the agent and environment naturally divides into segments called *episodes*. Each episode concludes with a terminal state, after which the agent either resets to a fixed starting state or draws from a standard distribution of initial states. Such tasks are referred to as *episodic tasks*.

In many RL scenarios, the interaction between the agent and environment continues indefinitely, without clear endpoints or episodes. These are known as *continuing tasks*. For such tasks, the return defined by a finite endpoint $T$ (as in Equation 3.13) is problematic because $T = \infty$ would make the return infinite, posing challenges for optimization.

To address this, a *discounting* approach is introduced, where future rewards are progressively reduced by a factor, reflecting their diminishing value over time. The agent then aims to maximize the expected *discounted return*:

$$G_t := R_{t+1} + \gamma R_{t+2} + \gamma^2 R_{t+3} + \cdots = \sum_{k=0}^{\infty} \gamma^k R_{t+k+1} \tag{3.14}$$

where $\gamma \in [0, 1]$ is the discount factor. This factor determines the present value of future rewards: if $\gamma = 0$, the agent focuses only on maximizing immediate rewards, disregarding long-term outcomes. As $\gamma$ approaches 1, the agent increasingly values future rewards. This formulation enables the agent to balance short-term gains with long-term objectives in continuing tasks.

#### 3.1.3.3 Policies and Value Functions

In RL, almost all algorithms involve estimating value functions, which assess the expected return for the agent being in a certain state, or taking a particular action in a state. This "goodness" measure, defined in terms of future rewards, depends on the agent's actions and is calculated with respect to specific strategies, known as policies.

A policy, denoted $\pi$, is essentially a mapping from states to probabilities of selecting each possible action. If the agent follows policy $\pi$ at time $t$, then $\pi(a|s)$ represents the probability of choosing action $a$ when in state $s$ at time $t$.

The value function for a state $s$ under a policy $\pi$, written as $v_\pi(s)$ , is the expected return the agent would achieve starting in state $s$ and continuing to act according to policy $\pi$. For MDPs, the value function $v_\pi$ can be formally defined as follows:

$$v_\pi(s) := \mathbb{E}_\pi\left[G_t|S_t=s\right] = \mathbb{E}_\pi\left[\sum_{k=0}^{\infty}\gamma^k R_{t+k+1}|S_t=s\right] \tag{3.15}$$

where $\mathbb{E}_\pi[\cdot]$ represents the expected value of a random variable given that the agent follows policy $\pi$, and $t$ denotes any time step. Importantly, the value of the terminal state, if one exists, is defined to be zero. This function $v_\pi$ is referred to as the state-value function for policy $\pi$, as it estimates the expected return starting from each state when the agent adheres to policy $\pi$ thereafter.

The function $q_\pi(s,a)$ known as the action-value function for policy $\pi$, represents the expected return starting from state $s$, taking action $a$, and subsequently following policy $\pi$. Formally, it is defined as:

$$q_\pi(s,a) := \mathbb{E}_\pi\left[G_t|S_t=s, A_t=a\right] = \mathbb{E}_\pi\left[\sum_{k=0}^{\infty}\gamma^k R_{t+k+1}|S_t=s, A_t=a\right] \tag{3.16}$$

This function provides an estimate of the total return expected when action $a$ is taken in state $s$, and all future actions are chosen according to $\pi$.

A fundamental property of value functions in RL is that they satisfy recursive relationships. For any policy $\pi$ and any state $s$, the value of $s$ is related to the values of its possible successor states. This relationship can be expressed through the following recursive equation:

$$v_\pi(s) = \sum_a \pi(a|s)\sum_{s',r} p(s',r|s,a)\left[r+\gamma v_\pi(s')\right] \tag{3.17}$$

Equation (3.17) represents the **Bellman equation** for $v_\pi$, which establishes a recursive relationship between the value of a state $v_\pi(s)$ and the values of its possible successor states $v_\pi(s')$.

#### 3.1.3.4 Optimal Policies and Optimal Value Functions

Solving a RL task involves identifying a policy that maximizes cumulative rewards over time. For finite MDPs, an optimal policy can be rigorously defined through value functions, which establish a partial ordering over policies. Specifically, a policy $\pi$ is considered better than or equal to another policy $\pi'$ if its expected return is greater than or equal to that of $\pi'$ for all states. Formally, $\pi \geq \pi'$ if and only if $v_\pi(s) \geq v_{\pi'}(s)$ for all $s \in S$. At least one policy exists that is superior to or equal to all others, known as an optimal policy. Although multiple optimal policies may exist, they are collectively denoted by $\pi_*$. All optimal policies share a common state-value function, referred to as the optimal state-value function, denoted $v_{\pi_*}$, and defined as:

$$v_{\pi_*}(s) := \max_\pi v_\pi(s) \tag{3.18}$$

for all $s \in S$.

Optimal policies also share the same optimal action-value function, denoted $q_{\pi_*}$, which is defined as:

$$q_{\pi_*}(s,a) := \max_\pi q_\pi(s,a) \tag{3.19}$$

for all $s \in S$ and $a \in A(s)$.

Since $v_{\pi_*}$ represents the value function for a policy, it must satisfy the self-consistency condition as expressed by the Bellman equation for state values (Equation 3.17). However, because $v_{\pi_*}$ is also the optimal value function, its consistency condition can be reformulated without referencing any specific policy. This reformulation is known as the Bellman optimality equation for $v_{\pi_*}$. Intuitively, the Bellman optimality equation indicates that the value of a state under an optimal policy corresponds to the expected return of the best possible action taken from that state:

$$v_{\pi_*}(s) = \max_a \sum_{s',r} p(s',r|s,a)\left[r + \gamma v_{\pi_*}(s')\right] \tag{3.20}$$

Once $v_{\pi_*}$ is known, determining an optimal policy becomes straightforward. For each state $s$, the optimal actions are those that maximize the Bellman optimality equation.

#### 3.1.3.5 Actor-critic Methods

Actor-critic RL combines the benefits of both policy-based and value-based learning through two distinct components: the *actor* and the *critic*. The actor's role is to learn a policy that maximizes the expected reward, while the critic evaluates the actions taken by the actor, learning a value function that provides feedback on the quality of those actions. This approach allows the actor to refine its policy based on both immediate and cumulative feedback. The key components of the actor-critic approach are discussed in detail as follows:

1. **Actor:** The actor is responsible for updating the policy used to select actions in each state. This policy, represented as $\pi(a|s;\theta)$ and parameterized by $\theta$, assigns probabilities to each action $a$ for a given state $s$. The actor's policy is updated based on feedback from the critic, enabling it to increase the probability of actions that are likely to yield higher rewards.
2. **Critic:** The critic network, represented as $V(s; w)$ and parameterized by $w$, estimates the value of the current policy by calculating a value function. It evaluates the actor's decisions using the temporal difference (TD) error, denoted as $\delta$. The TD error is calculated as:

   $$\delta_t := R_{t+1} + \gamma V(s_{t+1}; w) - V(s_t; w) \tag{3.21}$$

   where $V(s)$ represents the estimated value of state $s$, $R_{t+1}$ is the observed reward, and $\gamma$ is the discount factor. The TD error represents the difference between the expected and observed returns, and it drives updates to the critic's value estimates via:

   $$w \leftarrow w + \beta \delta_t \nabla_w V(s; w) \tag{3.22}$$

where $\nabla_w V(s; w)$ is the gradient of the critic network with respect to $w$, and $\beta$ is the learning rate.

#### 3.1.3.5.1 *Actor Update Using the Policy Gradient*

The actor leverages the TD error (Equation 3.21) to guide policy adjustments using the policy gradient. This enables the actor network to increase its probability of favorable actions. The update rule for the actor's policy parameters is:

$$\theta \leftarrow \theta + \alpha\delta_t \nabla_\theta \log \pi_\theta(a|s) \tag{3.23}$$

where $\nabla_\theta \log \pi_\theta(a|s)$ is the gradient of the logarithm of the actor network with respect to $\theta$ and $\alpha$ is the learning rate.

#### 3.1.3.5.2 *Proximal Policy Optimization*

Proximal policy optimization (PPO) [73] is an RL algorithm that combines aspects of policy optimization and value-based methods, in an actor-critic framework. PPO stabilizes learning by limiting the size of policy updates. This prevents drastic changes of the policy. PPO optimizes a *clipped objective function*, which ensures that updates remain within a predefined range. The core components of the PPO algorithm are defined as follows:

- **Objective Function:** PPO computes a probability ratio between the current and previous policy for a given action $a$ in state $s$:

$$\hat{r}_t(\theta) := \frac{\pi_\theta(a|s)}{\pi_{\theta_{\text{old}}}(a|s)} \tag{3.24}$$

  The objective function is modified using a clipping mechanism to limit the size of policy updates.
- **Clipping Mechanism:** The clipped loss function balances exploration and exploitation while preserving stability in training:

$$L^{\text{CLIP}}(\theta) := \mathbb{E}\left[\min\left(\hat{r}_t(\theta)\widehat{A}_t, \text{clip}(\hat{r}_t(\theta), 1-\epsilon, 1+\epsilon)\widehat{A}_t\right)\right] \tag{3.25}$$

  where $\epsilon$ is a hyperparameter controlling the clipping range, and $\widehat{A}_t$ represents the estimated *advantage* function, indicating the relative value of an action.
- ***Advantage* Estimation:** PPO uses a separate critic network to estimate the *advantage* function, which provides feedback on the quality of an action.
- **Actor-critic Framework:** PPO utilizes both actor and critic networks. The actor network updates the policy by maximizing the clipped objective function, while the critic network estimates the value function.

### 3.1.4 Multi-agent Reinforcement Learning

Multi-agent RL (MARL) involves training multiple agents in a shared environment to develop optimal policies. Within a multi-agent system, agents interact with each other and their surroundings, working to achieve their goals. Similar to single-agent RL, MARL policies are refined through a trial-and-error process aimed at maximizing cumulative rewards. Figure 3.4 shows a simple schematic of a MARL system. In MARL, each of the $n$ agents

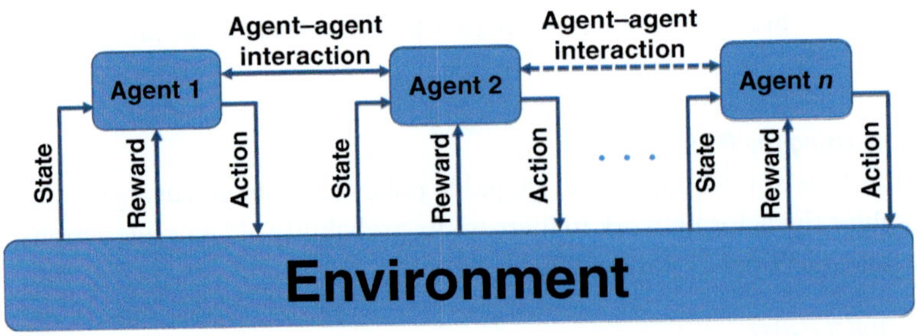


**Figure 3.4** A schematic diagram of multi-agent reinforcement learning. A set of $n$ agents receives individual information about the state of the environment, and choose actions to modify the state of the environment.

selects an individual action, collectively forming a joint action, which alters the environment's state. Agents receive individual rewards and observations, and this cycle continues until a predefined end condition is met.

The Markov game, or stochastic game (SG), generalizes the MDP framework to accommodate multiple agents in MARL settings.

Definition 1: A fully observable SG is characterized by the tuple $\langle N, S_i, A_i, p, r_i \rangle$, where $i \in N$ and $N = 1, 2, \ldots, n$ denotes the set of $n$ agents. Each agent $i$ has its own state space $S_i$ and action space $A_i$. The overall joint action space is given by the Cartesian product $A = A_1 \times A_2 \times \cdots \times A_n$. The transition dynamics are described by the probability function $p : S \times A \times S \rightarrow [0, 1]$, indicating the likelihood of transitioning between states given a joint action. For each agent $i \in N$, the reward function is defined as $r_i : S \times A \times S \rightarrow \mathbb{R}$.

MARL offers substantial advantages in specific cases over single-agent RL. One major benefit is its ability to decompose a complex, large-scale problem into smaller, more manageable sub-problems, with each agent responsible for part of the decision-making. This setting enables agents to learn cooperatively or competitively. An application of MARL in irrigation scheduling could involve multiple agents, each responsible for managing irrigation in specific regions across a large, heterogeneous field. In this MARL framework, individual agents may be assigned to manage irrigation for specific regions, relying on localized inputs such as soil moisture levels, crop requirements, and environmental factors. To promote coordination across the entire field, a centralized agent can be incorporated to align local decisions with broader objectives, such as maximizing crop yield and minimizing water consumption. Chapters 7–9 delve into how MARL can be leveraged to tackle the complexities of irrigation scheduling in large-scale agricultural settings.

## 3.2 Model Predictive Control

MPC is a control strategy that is aimed at optimizing system performance while accounting for constraints. Consequently, MPC is particularly suited to complex, multivariable control applications commonly encountered in industrial settings. At each time step, MPC formulates an optimization problem using the current system state to generate a sequence of control actions over a specified horizon. Only the first action in this sequence is implemented, with the process repeating as new state information becomes available. The typical design of MPC, as illustrated in Figure 3.5, includes the following components:

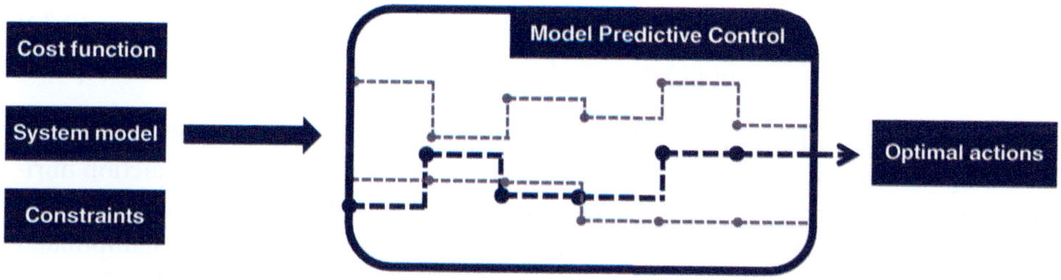


**Figure 3.5** An illustration of model predictive control.

1. **System Model:** The model must accurately capture the system's response to expected future inputs.
2. **Constraints:** One of the primary advantages of MPC lies in its capacity to directly incorporate system constraints, which are an unavoidable aspect of physical processes. These constraints must be expressed in mathematical form to be integrated into the control framework. Generally, constraints are specific to the problem at hand and should be selected accordingly. Certain constraints, such as those on manipulated inputs, are often dictated by the physical limitations of equipment and therefore cannot be violated. These are known as *hard constraints* and must be strictly observed in MPC optimization. For instance, in irrigation scheduling, the maximum irrigation amount allowed by a sprinkler sets a hard limit that cannot be relaxed. Other constraints, however, are desirable but more flexible, allowing for limited violations if necessary. These are termed *soft constraints* and can be incorporated into the MPC performance measure through the use of *slack variables*. For example, in irrigation scheduling, maintaining soil moisture content within a target range is desirable, yet minor violations of this range can be permitted.
3. **Performance Measure:** MPC belongs to a broader class of optimal control techniques that seek to optimize a defined performance criterion. This criterion typically captures trade-offs between multiple goals. In the context of irrigation scheduling, such goals might include reducing water consumption, lowering energy usage, and enhancing crop productivity.

### 3.2.1 Formulation of MPC

MPC is an online optimization-based strategy that determines control actions by minimizing a cost function over a future prediction horizon. It relies on a model of the system to forecast future behavior and is cast as a dynamic optimization problem. At a given sampling instant $k$, the MPC problem can be expressed in the following general form:

$$\min_{u} \quad J_N(x, u) \tag{3.26}$$

$$\text{s.t.} \quad x(j+1|k) = F(x(j|k), u(j|k), w(j|k)), \qquad j = k, \ldots, k+N-1 \tag{3.27}$$

$$y(j|k) = H(x(j|k)),\ j = k, \ldots, k+N \tag{3.28}$$

$$x(k|k) = \hat{x}(k) \tag{3.29}$$

$$x(j|k) \in X, y(j|k) \in Y, \; j = k, \ldots, k+N \tag{3.30}$$

$$u(j|k) \in U, \; j = k, \ldots, k+N-1 \tag{3.31}$$

In this formulation, $J_N$ is the performance measure (or cost function) to be minimized, representing the system's performance from time $k$ to $k+N$, where $N$ is the prediction horizon of the MPC. The term $x(j|k)$ denotes the state variable at time $j$ as predicted in the MPC framework based on information available at time $k$. Similarly, $u(j|k)$ is the manipulated input generated by MPC for time $j$, and $w(j|k)$ is the estimated value used in MPC for the uncontrolled input $w$ at time $j$. Equation (3.27) represents the system model, describing the system's evolution over time, while Equation (3.28) is the output equation that defines the relationship between the output and the state. Equation (3.29) sets the initial condition of the system within the MPC optimization. Equation (3.30) represents constraints on the state and output, and Equation (3.31) specifies the constraints on the manipulated input. In the MPC optimization problem, the decision variable is $\boldsymbol{u} = \{u(j|k)\}_k^{k+N-1}$, which is determined such that $J_N$ is minimized while satisfying the system model and constraints outlined in Equations (3.27)–(3.31).

After solving the MPC optimization problem, a trajectory of the optimized manipulated input $\boldsymbol{u}^* = \{u^*(j|k)\}_k^{k+N-1}$ is obtained. However, only the first value $u^*(k|k)$ is actually applied to the system; that is, $u(k) = u^*(k|k)$. The remaining optimized input trajectory is discarded. At the next time step $k+1$, the MPC optimization problem is updated with a new initial condition based on system measurement or estimation and is solved again. This process repeats at each sampling time. This approach is called *receding horizon implementation*. It incorporates feedback information from the system at every sampling time, which is essential for maintaining the stability and optimality of MPC.

### 3.2.2 Solving MPC

The MPC optimization problem in Equations (3.26)–(3.31) needs to be solved at each sampling time to incorporate the most recent information about the system. The MPC optimization problem is a NLP. Thus, open-source NLP solvers, such as the interior point optimizer Ipopt [74], can be used to solve the resulting NLP. Most NLP solvers seek local optimal solutions and require the user to provide an initial guess of the solution. A well-chosen initial guess can greatly aid in finding the optimal solution. In some MPC formulations, the decision variables may include both integer-valued and continuous variables. In this case, the resulting optimization problem becomes a mixed-integer NLP, and solvers such as Gurobi [75], CPLEX [76], or BONMIN [77] can be employed to address these problems.

### 3.2.3 Feasibility of MPC

When incorporating constraints in MPC design, it is essential to ensure that the optimization problem remains feasible—meaning a solution exists that satisfies all specified constraints. To manage cases of infeasibility, certain hard constraints may be reformulated as soft constraints, thereby allowing limited flexibility when strict compliance cannot be achieved.

### 3.2.4 Tuning of MPC

MPC requires careful tuning to achieve optimal performance. Key tuning parameters include the prediction horizon and the weighting factors for various terms within the cost function. For instance, the prediction horizon must be sufficiently long to ensure that the current MPC action accounts for future system behavior, thereby optimizing overall control performance. Similarly, the weighting factors within the performance measure should be precisely adjusted to ensure the MPC meets desired performance standards. Simulation experiments can be conducted to fine-tune these weighting factors.

## 3.3 Conclusion

In this chapter, machine learning and MPC were reviewed. For machine learning, the three primary paradigms were discussed, along with specific techniques such as feedforward neural networks, long short-term memory networks, PINNs, $k$-means clustering, actor-critic RL methods, proximal policy optimization, and MARL. This was followed by a discussion on MPC, covering key topics related to its design, formulation, solution methods, feasibility, and tuning.

# 4

# Soil Moisture and Hydraulic Parameter Estimation in Agro-hydrological Systems

## 4.1 Introduction

Reliable soil moisture data are a vital input for closed-loop irrigation scheduling systems. Among the various sensing methods available, microwave remote sensing offers a practical solution for capturing spatially distributed soil moisture over large agricultural areas. In fields equipped with center pivot irrigation systems, microwave sensors can be installed directly on the irrigation equipment. As the system rotates, these sensors collect soil moisture data, which is subsequently processed to produce spatial maps of soil moisture.

Despite its usefulness, microwave remote sensing is limited in its ability to provide complete temporal and spatial coverage. These limitations can introduce gaps in the data, which in turn affect the performance of model-based irrigation control. To address this, sequential data assimilation techniques are used to integrate limited sensor data with agro-hydrological models, providing a more complete and continuous estimate of soil moisture.

The accuracy of soil moisture estimation using agro-hydrological models is dependent on soil hydraulic parameters. These parameters define the soil's water retention and conductivity properties and are essential for simulating water movement. They are also valuable for delineating management zones, which support localized irrigation control. In most practical cases, the true values of these parameters are unknown and must be inferred from observed data. Joint estimation of soil moisture and hydraulic parameters is possible through data assimilation, leveraging the coupling between soil moisture state variables and parameters.

Before performing parameter estimation, it is necessary to assess which parameters can be uniquely estimated from the available measurement data. This process, known as parameter estimability analysis, helps determine the subset of parameters that can be estimated with confidence. In cases where only a few parameters are estimable due to limited or noisy measurement data, parameter selection is applied to focus estimation efforts on this subset, improving both accuracy and computational efficiency.

This chapter presents a method for the simultaneous estimation of soil moisture and selected soil hydraulic parameters in large-scale agricultural fields. The approach uses soil moisture data collected by microwave sensors mounted on a center pivot irrigation system. The Richards' equation, formulated in cylindrical coordinates to align with the field geometry, is used to model soil water movement. A sensitivity analysis is conducted to evaluate

*Precision Irrigation for Agriculture: Integrating Machine Learning and Optimal Control Strategies*, First Edition. Bernard Twum Agyeman and Jinfeng Liu.

parameter estimability, followed by parameter selection using the orthogonal projection technique. The extended Kalman filter (EKF) is used to assimilate sensor data and estimate both soil moisture and selected parameters. For parameters not selected for estimation, Kriging interpolation is used to update their values. The method is evaluated through simulations and a real-world case study, showing improved estimation accuracy and reduced computational load. This work builds on earlier studies reported in [78, 79, 80].

## 4.2 Model Development

This chapter examines an agro-hydrological system equipped with a center pivot irrigation system, as shown in Figure 4.1. The field model utilized is the 3D form of the Richards' equation in cylindrical coordinates, which was first introduced in Chapter 2. The cylindrical coordinate system is particularly advantageous for this scenario because it accounts for the rotational movement of the center pivot, allowing for a more accurate depiction of water distribution. Furthermore, this system is well-suited for managing spatial data gathered from microwave remote sensors. The Richards' equation in cylindrical coordinates is defined as:

$$C(\psi)\frac{\partial \psi}{\partial t} = \frac{1}{r}\frac{\partial}{\partial r}\left[rK(\psi)\frac{\partial \psi}{\partial r}\right] + \frac{1}{r}\frac{\partial}{\partial \theta}\left[\frac{K(\psi)}{r}\frac{\partial \psi}{\partial \theta}\right] + \frac{\partial}{\partial z}\left[K(\psi)\left(\frac{\partial \psi}{\partial z}+1\right)\right] - S(\psi, z) \tag{4.1}$$

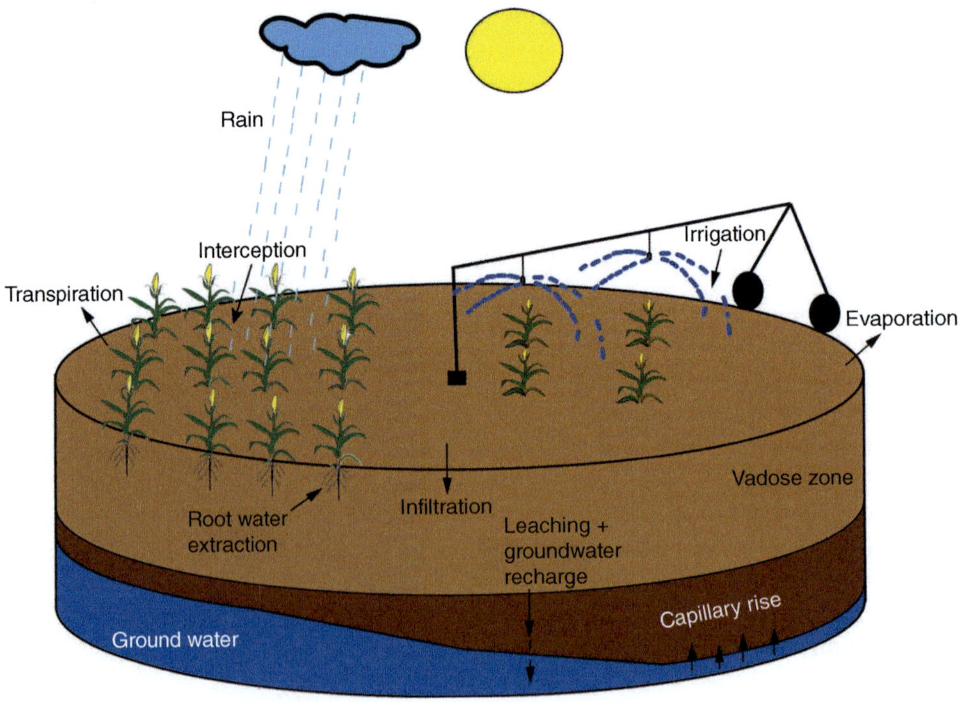


**Figure 4.1** Agro-hydrological system with center pivot irrigation. *Source*: [81] with permission of Elsevier.

To numerically solve Equation (4.1) [81], the central difference scheme is used to approximate the spatial derivatives in the radial ($r$), azimuthal ($\theta$), and axial ($z$) directions. After approximating these spatial derivatives, the equation is then converted into an ordinary differential equation (ODE) with respect to time, which is expressed as:

$$
\begin{aligned}
\frac{d\psi}{dt} = \frac{1}{C_{e_r,e_\theta,k}(\psi)}\Bigg[&\left(\frac{1}{r_{e_r,e_\theta,k}\Delta r_i}\left[r_{e_r+\frac{1}{2},e_\theta,k}K_{e_r+\frac{1}{2},e_\theta,k}(\psi)\left(\frac{\psi_{e_r+1,e_\theta,k}-\psi_{e_r,e_\theta,k}}{\Delta r_E}\right)\right.\right.\\
&\left.\left.-r_{e_r-\frac{1}{2},e_\theta,k}K_{e_r-\frac{1}{2},e_\theta,k}(\psi)\left(\frac{\psi_{e_r,e_\theta,k}-\psi_{e_r-1,e_\theta,k}}{\Delta r_W}\right)\right]\right)\\
&+\left(\frac{1}{r_{e_r,e_\theta,k}\Delta\theta_j}\left[\frac{K_{e_r,e_\theta+\frac{1}{2},k}(\psi)}{r_{e_r,e_\theta+\frac{1}{2},k}}\left(\frac{\psi_{e_r,e_\theta+1,k}-\psi_{e_r,e_\theta,k}}{\Delta\theta_T}\right)\right.\right.\\
&\left.\left.-\frac{K_{e_r,e_\theta-\frac{1}{2},k}(\psi)}{r_{e_r,e_\theta-\frac{1}{2},k}}\left(\frac{\psi_{e_r,e_\theta,k}-\psi_{e_r,e_\theta-1,k}}{\Delta\theta_D}\right)\right]\right)\\
&+\left(\frac{1}{\Delta z_k}\left[K_{e_r,e_\theta,k+\frac{1}{2}}(\psi)\left(\frac{\psi_{e_r,e_\theta,k+1}-\psi_{e_r,e_\theta,k}}{\Delta z_N}+1\right)\right.\right.\\
&\left.\left.-K_{e_r,e_\theta,k-\frac{1}{2}}(\psi)\left(\frac{\psi_{e_r,e_\theta,k}-\psi_{e_r,e_\theta,k-1}}{\Delta z_S}+1\right)\right]\right)-S\left(\psi_{e_r,e_\theta,k},z\right)\Bigg]
\end{aligned}
\tag{4.2}
$$

In the above ODE, $e_r \in [0, N_r+1]$, $e_\theta \in [0, N_\theta+1]$, and $k \in [0, N_z+1]$ represent the position indices in the $r$, $\theta$, and $z$ directions, respectively, with $e_r$, $e_\theta$, $k \in \mathbb{Z}_+$. The term $N_a$ in Equation (4.2) denotes the total number of nodes in the $a$th direction, where $a \in r, \theta, z$. The spatial discretization terms are defined as follows: $\Delta r_E = r_{e_r+1,e_\theta,k} - r_{e_r,e_\theta,k}$, $\Delta r_W = r_{e_r,e_\theta,k} - r_{e_r-1,e_\theta,k}$, $\Delta r_i = \frac{1}{2}(\Delta r_E + \Delta r_W)$. $r_{e_r\pm\frac{1}{2},e_\theta,k} = \frac{1}{2}(r_{e_r,e_\theta,k} + r_{e_r\pm1,e_\theta,k})$, $\Delta\theta_T = \theta_{e_r,e_\theta+1,k} - \theta_{e_r,e_\theta,k}$, $\Delta\theta_D = \theta_{e_r,e_\theta,k} - \theta_{e_r,e_\theta-1,k}$ and $\Delta\theta_j = \frac{1}{2}(\Delta\theta_T + \Delta\theta_D)$. $r_{e_r,e_\theta\pm\frac{1}{2},k} = \frac{1}{2}(r_{e_r,e_\theta,k} + r_{e_r,e_\theta\pm1,k}) = r_{e_r,e_\theta,k}$, $\Delta z_N = z_{e_r,e_\theta,k+1} - z_{e_r,e_\theta,k}$, $\Delta z_S = z_{e_r,e_\theta,k} - z_{e_r,e_\theta,k-1}$, and $\Delta z_k = \frac{1}{2}(\Delta z_N + \Delta z_S)$. Note that $K(\psi)$ and $C(\psi)$ in Equation (4.1) refer to the hydraulic conductivity (i.e., Equation 2.22) and the specific moisture capacity (i.e., Equation 2.23) as introduced in Chapter 2. The hydraulic conductivities in each spatial direction are approximated as follows: $K_{e_r\pm\frac{1}{2},e_\theta,k}(\psi) = \frac{1}{2}\left(K\left(\psi_{e_r,e_\theta,k}\right) + K\left(\psi_{e_r\pm1,e_\theta,k}\right)\right)$, $K_{e_r,e_\theta\pm\frac{1}{2},k}(\psi) = \frac{1}{2}\left(K\left(\psi_{e_r,e_\theta,k}\right) + K\left(\psi_{e_r,e_\theta\pm1,k}\right)\right)$ and $K_{e_r,e_\theta\pm\frac{1}{2},k}(\psi) = \frac{1}{2}\left(K\left(\psi_{e_r,e_\theta,k}\right) + K\left(\psi_{e_r,e_\theta\pm1,k}\right)\right)$. It is worth noting that the nodes in the $r$, $\theta$, and $z$ directions are centered at the $e_r$th, $e_\theta$th, and $k$th compartments, respectively.

From Equation (4.1), it is evident that the radius $r$ appears in the denominator of the spatial derivatives with respect to the radial and azimuthal directions. Consequently, as $r$ approaches zero at the axis of the field, these terms become unbounded, resulting in a singularity at $r = 0$. To address this, L'Hôpital's rule is applied to the first and second terms on the right-hand side of Equation (4.1), resulting in the following expressions at $r = 0$:

$$
\lim_{r\to0}\frac{1}{r}\frac{\partial}{\partial r}\left[rK(\psi)\frac{\partial\psi}{\partial r}\right] = \lim_{r\to0}\frac{\frac{\partial}{\partial r}\left(\frac{\partial}{\partial r}\left[rK(\psi)\frac{\partial\psi}{\partial r}\right]\right)}{\frac{\partial}{\partial r}(r)} = 2\frac{\partial}{\partial r}\left(K(\psi)\frac{\partial\psi}{\partial r}\right)
\tag{4.3}
$$

$$\lim_{r\to 0}\frac{1}{r}\frac{\partial}{\partial\theta}\left[\frac{K(\psi)}{r}\frac{\partial\psi}{\partial\theta}\right] = \lim_{r\to 0}\frac{\frac{\partial}{\partial r}\left(\frac{\partial}{\partial\theta}\left[K(\psi)\frac{\partial\psi}{\partial\theta}\right]\right)}{\frac{\partial}{\partial r}(r^2)}$$

$$= \frac{1}{2}\frac{\partial}{\partial r}\left(\frac{\partial}{\partial r}\left(\frac{\partial}{\partial\theta}\left[K(\psi)\frac{\partial\psi}{\partial\theta}\right]\right)\right) \tag{4.4}$$

Substituting Equations (4.3) and (4.4) into Equation (4.1) yields the cylindrical coordinate version of the Richards' equation at $r=0$, expressed as:

$$C(\psi)\frac{\partial\psi}{\partial t} = 2\frac{\partial}{\partial r}\left(K(\psi)\frac{\partial\psi}{\partial r}\right) + \frac{1}{2}\frac{\partial}{\partial r}\left(\frac{\partial}{\partial r}\left(\frac{\partial}{\partial\theta}[K(\psi)\frac{\partial\psi}{\partial\theta}]\right)\right)$$

$$+ \frac{\partial}{\partial z}[K(\psi)(\frac{\partial\psi}{\partial z}+1)] - S(\psi,z) \tag{4.5}$$

The spatial derivative approximation in the axial direction for $r\neq 0$, as outlined in Equation (4.2), holds true at the axis ($r=0$). However, to handle the spatial derivatives at the axis, central difference approximations must be derived for the first and second terms in Equation (4.5). The central difference approximations for these terms are as follows:

$$2\frac{\partial}{\partial r}\left(K(\psi)\frac{\partial\psi}{\partial r}\right) \approx \frac{2}{\Delta r_i^2}\left(K_{e_r+\frac{1}{2},e_\theta,k}(\psi)\left[\psi_{e_r+1,e_\theta,k}-\psi_{e_r,e_\theta,k}\right]\right.$$

$$\left.-K_{e_r-\frac{1}{2},e_\theta,k}(\psi)\left[\psi_{e_r,e_\theta,k}-\psi_{e_r-1,e_\theta,k}\right]\right) \tag{4.6}$$

$$\frac{1}{2}\frac{\partial}{\partial r}\left(\frac{\partial}{\partial r}\left(\frac{\partial}{\partial\theta}\left[K(\psi)\frac{\partial\psi}{\partial\theta}\right]\right)\right) \approx \frac{1}{2\Delta r_i^2\Delta\theta_k^2}\left[\left(K_{e_r+1,e_\theta+\frac{1}{2},k}(\psi)\left[\psi_{e_r+1,e_\theta+1,k}-\psi_{e_r+1,e_\theta,k}\right]\right.\right.$$

$$\left.-K_{e_r+1,e_\theta-\frac{1}{2},k}(\psi)\left[\psi_{e_r+1,e_\theta,k}-\psi_{e_r+1,e_\theta-1,k}\right]\right)$$

$$-2\left(K_{e_r,e_\theta+\frac{1}{2},k}(\psi)\left[\psi_{e_r,e_\theta+1,k}-\psi_{e_r,e_\theta,k}\right]\right.$$

$$\left.-K_{e_r,e_\theta-\frac{1}{2},k}(\psi)\left[\psi_{e_r,e_\theta,k}-\psi_{e_r,e_\theta-1,k}\right]\right)$$

$$+\left(K_{e_r-1,e_\theta+\frac{1}{2},k}(\psi)\left[\psi_{e_r-1,e_\theta+1,k}-\psi_{e_r-1,e_\theta,k}\right]\right.$$

$$\left.\left.-K_{e_r-1,e_\theta-\frac{1}{2},k}(\psi)\left[\psi_{e_r-1,e_\theta,k}-\psi_{e_r-1,e_\theta-1,k}\right]\right)\right] \tag{4.7}$$

The numerical solution of Equation (4.1) is carried out subject to the following boundary conditions:

$$\frac{\partial\psi(r,\theta,z)}{\partial r} = 0 \quad \text{at } (r=0,\ \theta,\ z) \tag{4.8}$$

$$\frac{\partial\psi(r,\theta,z)}{\partial r} = 0 \quad \text{at } (r=H_r,\ \theta,\ z) \tag{4.9}$$

$$\frac{\partial(\psi(r,\theta,z)+z)}{\partial z}=1 \qquad \text{at } (r,\ \theta,\ z=0) \tag{4.10}$$

$$\frac{\partial(\psi(r,\theta,z))}{\partial z}=-1-\frac{u^{\text{irr}}-\text{EV}}{K(\psi)} \qquad \text{at } (r,\ \theta,\ z=H_z) \tag{4.11}$$

$$\psi(r,\ \theta=0,\ z,\ t)=\psi(r,\ \theta=2\pi,\ z,\ t) \tag{4.12}$$

In Equation (4.9), $H_r$ refers to the overall radius of the field, while $H_z$ in Equation (4.11) specifies the depth of the soil column. The terms $u^{\text{irr}}$ (m/s) and EV (m/s) in Equation (4.11) denote the irrigation amount and the potential evapotranspiration, respectively.

Following numerical discretization, the field model can be represented in state-space form as follows:

$$\dot{x}(t)=\mathcal{F}(x(t),u(t),\ \mathbf{\Phi})+\omega(t) \tag{4.13}$$

$$y(t)=M(t)\mathcal{H}(x(t),\ \mathbf{\Phi})+v(t) \tag{4.14}$$

In Equation (4.13), $x(t)\in\mathbb{R}^{N_x}$ represents the state of the system, comprising of $N_x$ pressure head values. The total number of discretization nodes, $N_x$, is calculated as $N_x=N_r\times N_\theta\times N_z$. The inputs (such as irrigation, crop coefficient, rainfall, and reference evapotranspiration) and model disturbances are denoted as $u(t)\in\mathbb{R}^{N_u}$ and $\omega(t)\in\mathbb{R}^{N_x}$, respectively. Soil heterogeneity in the field model is captured by assigning unique hydraulic parameters $\{K_s,\theta_s,\theta_r,\alpha,n\}$ to each spatial node. In Equations (4.13) and (4.14), $\mathbf{\Phi}\in\mathbb{R}^{N_p}$ represents the full set of hydraulic parameters for all $N_x$ nodes.

In Equation (4.14), the vectors $y(t)\in\mathbb{R}^{N_y}$ and $v(t)\in\mathbb{R}^{N_y}$ represent the measurement output and the associated noise, respectively. This equation serves as a generalized version of Equation (2.9), reflecting the fact that microwave sensors measure volumetric soil moisture. The matrix $M(t)$ acts as a selection matrix, specifying which nodes in the soil profile contribute to the sensor measurements at time $t$, based on the sensor's penetration depth. Since the center pivot moves while irrigating the field, the spatial locations of the observations vary with time. Consequently, $M(t)$ changes dynamically to reflect the shifting measurement positions. The matrix $M(t)$ is expressed as follows:

$$M(t)=\frac{1}{N_c}\left[I_{N_y\times N_y}(t)\quad \emptyset_{N_y\times(N_x-N_y)}(t)\right] \tag{4.15}$$

where $N_c$ indicates the number of axial nodes falling within the penetration depth of the sensor, while $N_y$ represents the total number of soil moisture measurements recorded at time $t$. The selection matrix $M(t)$ is composed of the identity matrix $I$ and the zero matrix $\emptyset$, whose positions shift over time. This variation reflects the dynamic nature of the measurement locations, which change with each time step as the center pivot rotates across the field.

## 4.3 Sensitivity Analysis for Parameter Estimability

Sensitivity analysis is essential for understanding how changes in model parameters influence system outputs and plays a central role in determining which parameters can be reliably estimated. The analysis involves computing sensitivity trajectories, which describe how outputs respond to variations in individual parameters. These trajectories are then organized into a matrix, referred to as the sensitivity matrix, which serves as the basis for assessing parameter estimability. For the agro-hydrological model considered in this study, the sensitivity of the system state $x$ with respect to the parameter vector $\mathbf{\Phi}$ is governed by the following set of equations:

$$\frac{dx_{\mathbf{\Phi}}(t)}{dt} = \frac{\partial \mathcal{F}}{\partial x} x_{\mathbf{\Phi}}(t) + \frac{\partial \mathcal{F}}{\partial \mathbf{\Phi}} \tag{4.16}$$

$$y_{\mathbf{\Phi}}(t) = \frac{\partial \mathcal{H}}{\partial x} x_{\mathbf{\Phi}}(t) + \frac{\partial \mathcal{H}}{\partial \mathbf{\Phi}} \tag{4.17}$$

where $x_{\mathbf{\Phi}}(t) = \frac{\partial x(t)}{\partial \mathbf{\Phi}}$ and $y_{\mathbf{\Phi}}(t) = \frac{\partial y(t)}{\partial \mathbf{\Phi}}$.

These sensitivity equations are solved alongside the main model equations. To perform the computation, one must provide an initial state, assign nominal parameter values, and set the initial sensitivity matrix to zero: $x_{\mathbf{\Phi}}(t_0) = \mathbf{0}_{N_x \times N_p}$. The integration proceeds over a set of time steps $t_0, t_1, \ldots, t_N$, with the total number of steps chosen to be at least equal to the number of parameters being analyzed. At each time step $t_k$, an instantaneous sensitivity matrix, $s_y(t_k, \mathbf{\Phi})$, is formed as:

$$s_y(t_k, \mathbf{\Phi}) = \begin{bmatrix} \frac{\partial y_1(t_k)}{\partial \Phi_1} & \frac{\partial y_1(t_k)}{\partial \Phi_2} & \cdots & \frac{\partial y_1(t_k)}{\partial \Phi_{N_p}} \\ \frac{\partial y_2(t_k)}{\partial \Phi_1} & \frac{\partial y_2(t_k)}{\partial \Phi_2} & \cdots & \frac{\partial y_2(t_k)}{\partial \Phi_{N_p}} \\ \vdots & \vdots & \vdots & \vdots \\ \frac{\partial y_{N_y}(t_k)}{\partial \Phi_1} & \frac{\partial y_{N_y}(t_k)}{\partial \Phi_2} & \cdots & \frac{\partial y_{N_y}(t_k)}{\partial \Phi_{N_p}} \end{bmatrix} \tag{4.18}$$

Since the elements in $s_y(t_k, \mathbf{\Phi})$ can differ in magnitude and physical units, direct comparison is often not meaningful. Therefore, normalization is applied to each element, resulting in a scaled instantaneous sensitivity matrix $\tilde{s}_y(t_k, \mathbf{\Phi})$:

$$\tilde{s}_y(t_k, \mathbf{\Phi}) = \begin{bmatrix} \frac{\Phi_1}{y_1(t_k)} \frac{\partial y_1(t_k)}{\partial \Phi_1} & \frac{\Phi_2}{y_1(t_k)} \frac{\partial y_1(t_k)}{\partial \Phi_2} & \cdots & \frac{\Phi_{N_p}}{y_1(t_k)} \frac{\partial y_1(t_k)}{\partial \Phi_{N_p}} \\ \frac{\Phi_1}{y_2(t_k)} \frac{\partial y_2(t_k)}{\partial \Phi_1} & \frac{\Phi_2}{y_2(t_k)} \frac{\partial y_2(t_k)}{\partial \Phi_2} & \cdots & \frac{\Phi_{N_p}}{y_2(t_k)} \frac{\partial y_2(t_k)}{\partial \Phi_{N_p}} \\ \vdots & \vdots & \vdots & \vdots \\ \frac{\Phi_1}{y_{N_y}(t_k)} \frac{\partial y_{N_y}(t_k)}{\partial \Phi_1} & \frac{\Phi_2}{y_{N_y}(t_k)} \frac{\partial y_{N_y}(t_k)}{\partial \Phi_2} & \cdots & \frac{\Phi_{N_p}}{y_{N_y}(t_k)} \frac{\partial y_{N_y}(t_k)}{\partial \Phi_{N_p}} \end{bmatrix} \tag{4.19}$$

where $y_i(t_k)$ denotes the $i$th observed value at time step $t_k$, while $\Phi_j$ corresponds to the nominal value of the $j$th parameter.

The overall scaled sensitivity matrix, denoted as $\tilde{S}_y(t_0, t_1, \ldots, t_N, \mathbf{\Phi})$, is constructed by stacking all the scaled instantaneous matrices across time:

$$\tilde{S}_y(t_0, t_1, \ldots, t_N, \mathbf{\Phi}) = \begin{bmatrix} \tilde{s}_y(t_0, \mathbf{\Phi}) \\ \tilde{s}_y(t_1, \mathbf{\Phi}) \\ \vdots \\ \tilde{s}_y(t_k, \mathbf{\Phi}) \\ \vdots \\ \tilde{s}_y(t_N, \mathbf{\Phi}) \end{bmatrix} \tag{4.20}$$

If this overall matrix has full column rank, then all parameters are considered estimable. A rank-deficient matrix, however, implies that some parameters cannot be uniquely estimated from the available measurement data. To evaluate the rank of $\tilde{S}_y(t_0, t_1, \ldots, t_N, \mathbf{\Phi})$, singular value decomposition (SVD) can be employed.

### 4.3.1 Output Sensitivity Matrix for Spatially Varying Measurements

In this framework, soil moisture observations are acquired using microwave sensors installed on a rotating center pivot irrigation system. As the pivot rotates during irrigation, the spatial locations of the measurements change over time. This behavior is illustrated in Figure 4.2, where the sensor continuously observes different parts of the field. Typically, the sensor collects data from areas that have not yet received irrigation. In the figure, red markers indicate measurement locations at a specific time $t_k$, while blue markers show the regions being irrigated. At the subsequent time step $t_{k+1}$, the pivot advances to a new position, resulting in updated measurement locations.

The changing spatial nature of the measurements presents a challenge when constructing a consistent overall scaled sensitivity matrix. Since the observed nodes differ across time steps, directly combining the instantaneous scaled sensitivity matrices would lead to inconsistencies. To preserve spatial consistency, it is necessary to group together only those sensitivity matrices that correspond to the same measurement locations.

Accordingly, distinct overall scaled sensitivity matrices must be assembled for each unique measurement location. In the field model shown in Figure 4.2, this results in a separate sensitivity matrix for each azimuthal node. For example, consider a discretization with 40 azimuthal nodes and multiple pivot rotations. For the azimuthal node indexed by $e_\theta = 1$, the overall scaled sensitivity matrix is constructed using data from time steps $t_k = 1, 41, 81, 121, \ldots$, as these correspond to the moments when the sensor observes this specific location. The matrix is formed as:

$$\tilde{S}_y^1(1, 41, 81, \ldots, \mathbf{\Phi}) = \begin{bmatrix} \tilde{s}_y(1, \mathbf{\Phi}) \\ \tilde{s}_y(41, \mathbf{\Phi}) \\ \tilde{s}_y(81, \mathbf{\Phi}) \\ \vdots \end{bmatrix} \tag{4.21}$$

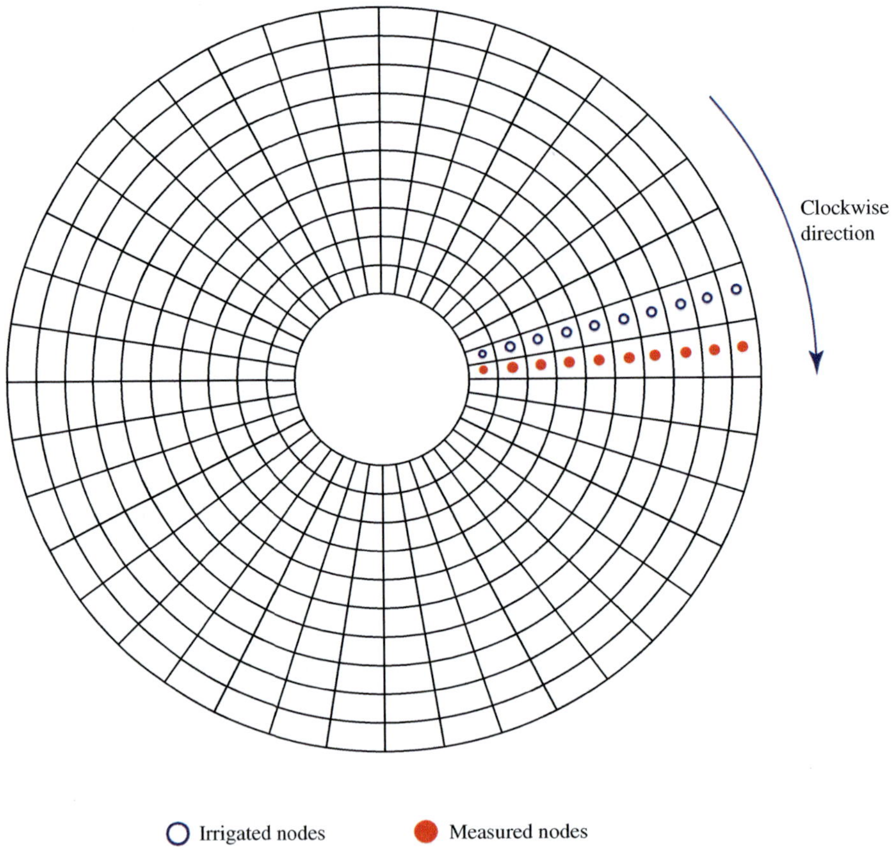


**Figure 4.2** Diagram illustrating microwave sensor integration on center pivot irrigation systems. *Source*: [79]/arXiv/CC BY 4.0.

For the second azimuthal node ($e_\theta = 2$), the matrix is constructed using time steps $t_k = 2, 42, 82, 122, \dots$, as follows:

$$\tilde{S}_y^2(2, 42, 82, \dots, \boldsymbol{\Phi}) = \begin{bmatrix} \tilde{s}_y(2, \boldsymbol{\Phi}) \\ \tilde{s}_y(42, \boldsymbol{\Phi}) \\ \tilde{s}_y(82, \boldsymbol{\Phi}) \\ \vdots \end{bmatrix} \tag{4.22}$$

This process is repeated for all other azimuthal nodes. SVD is then applied to each location-specific matrix to evaluate the estimability of the parameters based on the corresponding measurements.

## 4.4 Parameter Selection Through Orthogonal Projection

When the overall scaled sensitivity matrix is rank-deficient, it indicates that not all parameters in $\mathbf{\Phi}$ can be reliably estimated from the available measurements. In such cases, estimating all parameters may reduce the accuracy of the estimation process. To address this, a reduced set of estimable parameters must be selected for estimation. This section adopts the orthogonal projection technique to perform parameter selection based on the information contained in the overall scaled sensitivity matrix. Unlike techniques such as principal component analysis, the orthogonal projection method preserves the physical meaning of the selected parameters while reducing the number of variables considered during estimation.

The procedure evaluates the columns of the scaled sensitivity matrix in terms of their magnitude and linear independence. Given $\tilde{S}_y^i(\cdot)$—the overall scaled sensitivity matrix corresponding to azimuthal node $i$—the steps for selecting estimable hydraulic parameters are summarized as follows [44]:

1. Compute the norm for each column in $\tilde{S}_y^i(\cdot)$, initialize $j = 1$ and choose the column $X_j$ that has the largest norm.
2. For the selected $X_j$, calculate the associated information $Z_j$ in $\tilde{S}_y^i(\cdot)$ and determine the residual matrix $R_j$.

$$Z_j := X_j(X_j^T X_j)^{-1} X_j^T \tilde{S}_y^i(\cdot) \tag{4.23}$$

$$R_j := \tilde{S}_y^i(\cdot) - Z_j \tag{4.24}$$

3. Calculate the norm of each column in $R_j$. Then, identify the column from $\tilde{S}_y^i(\cdot)$ that corresponds to the column with the largest norm in $R_j$, and add it to the set of columns in $X_j$, creating $X_{j+1}$.
4. If $\mathrm{rank}(X_j) = \mathrm{rank}(\tilde{S}_y^i(\cdot))$ or if the largest column norm in $R_j$ falls below a specified threshold, end the process. The selected parameters of $\mathbf{\Phi}$ are those corresponding to the columns in $X_j$. If these conditions are not met, increment $j$ and repeat Steps 2–4.

This process is independently applied to each azimuthal node to determine the set of estimable parameters relevant to that location. To enable efficient online estimation, the parameter selection procedure can be performed offline using simulation-based sensitivity matrices. Notably, the set of estimable parameters may vary between azimuthal nodes. For instance, the parameters identified for estimation when the sensor observes node $i$ may differ from those associated with node $j$, where $i \neq j$. After completing the selection procedure for all nodes, the estimable parameters across the field are aggregated into a vector denoted by $\mathbf{\Phi}^e$. The remaining parameters, which are not selected for estimation, are grouped into $\mathbf{\Phi}^{ne}$.

## 4.5 EKF Design

To facilitate joint estimation of soil moisture and hydraulic parameters, the original state vector of the field model is augmented by including the selected estimable parameters, resulting in an augmented state defined as $x_a(t) = [x(t) \ \ \mathbf{\Phi}^e]$. This continuous-time formulation is subsequently discretized to obtain a discrete-time representation that also incorporates the non-estimable parameters. The discrete-time model is given by:

$$x_a(k+1) = \mathcal{F}_a(x_a(k), u(k), \mathbf{\Phi}^{ne}) + \omega_a(k) \tag{4.25}$$

$$y(k) = M_a(k)\mathcal{H}_a(x_a(k), \mathbf{\Phi}^{ne}) + v(k) \tag{4.26}$$

In this formulation, the non-estimable parameters in $\mathbf{\Phi}^{ne}$ are kept constant at their nominal values during the estimation process. After each sampling instance, the estimable parameters $\mathbf{\Phi}^e$ are updated, and the remaining parameters are refined using Kriging interpolation to capture spatial variation.

To perform data assimilation, this study employs the EKF. The EKF is initialized with an estimated augmented state, $\hat{x}_a(0)$, and an associated covariance matrix, $P_a(0|0)$.

In the prediction phase, the field model is used to propagate both the augmented state and the error covariance forward in time as follows:

$$\hat{x}_a(k+1|k) = \mathcal{F}_a(\hat{x}_a(k|k), u(k), \ \mathbf{\Phi}^{ne}) \tag{4.27}$$

$$P_a(k+1|k) = A(k)P_a(k|k)A(k)^T + Q \tag{4.28}$$

where $A(k) = \frac{\partial \mathcal{F}a}{\partial x_a}|_{\hat{x}_a(k|k),u(k)}$ and $Q$ represents the covariance matrix of the process uncertainty $\omega_a$.

In the update phase, the forecasted augmented state and its covariance matrix are corrected using the measurement $y(k+1)$ obtained at time step $k+1$, as follows:

$$\begin{aligned}\hat{x}_a(k+1|k+1) = \hat{x}_a(k+1|k) + G(k+1)[\,y(k+1) \\ - \mathcal{M}_a(k)\mathcal{H}_a(\hat{x}_a(k+1|k), \theta^{ne})]\end{aligned} \tag{4.29}$$

$$P_a(k+1|k+1) = [I - G(k+1)C(k+1)]P_a(k+1|k) \tag{4.30}$$

The Kalman gain is represented as $G(k+1)$ and it is computed as:

$$G(k+1) = P_a(k+1|k)C^T(k+1)[C(k+1)P_a(k+1|k)C^T(k+1) + R]^{-1} \tag{4.31}$$

where $C(k+1) = M_a(k)\frac{\partial \mathcal{H}a}{\partial x_a}|_{\hat{x}_a(k+1|k)}$ and $R$ represents the covariance matrix of the measurement noise $v$.

As previously described, the estimable parameter set $\mathbf{\Phi}^e$ are likely to change with time, given the spatial varying nature of the observations. If a particular parameter in $\mathbf{\Phi}^e$ is not active at a given time step, the relevant rows and columns in the matrices $A(k)$ and $C(k)$ are set to zero to exclude its influence from the update. This approach allows the EKF to adaptively manage time-varying estimability without the overhead of maintaining multiple covariance matrices, leading to improved computational efficiency.

## 4.6 Simulated Case Study

A simulated case study, designed to replicate the deployment of microwave radiometers, is initially used to evaluate the performance of the estimation strategy. The test field considered in this analysis spans a radial distance of 290 m and reaches a vertical distance of 0.3 m. It is divided into $N_x = 12{,}000$ spatial nodes, comprising 30 uniformly spaced segments along the radial axis, 40 in the azimuthal direction, and 10 along the depth. Temporal discretization is carried out with a resolution of 6 minutes. The simulation is performed over a heterogeneous field, and the spatial distribution of the nominal hydraulic parameters is depicted in Figure B.1 in Appendix B.1, where additional details on the derivation of these parameters are also provided.

The pressure head values at all nodes within the field model were randomly initialized within the range of −1.5 to −1.35 m. A center pivot system, rotating at a speed of 0.011 m/s, delivered a constant irrigation rate of 3.6 mm/day during the first 8 hours of each day. The field was assumed to be cultivated with barley in its developmental phase, with crop coefficient values ranging from 0.75 to 0.96. Daily reference evapotranspiration values were specified as 1.5, 1.90, 0.6, 0.8, and 2.40 mm/day. Both process and measurement noise were incorporated into the simulation, each with a mean of zero and standard deviations of $1 \times 10^{-6}$ m and $1 \times 10^{-2}$ $\text{m}^3/\text{m}^3$, respectively. At each time step, thirty measurements were used to update the elements of the augmented system during the EKF correction phase. The estimator was initialized with $\hat{x}_a(0) \in [1.10x_a(0), 1.15x_a(0)]$, where $\hat{x}_a(0)$ denotes the initial estimate of the state and $x_a(0)$ represents the true initial state.

The simulation adopted a sensing depth of 0.1 m, which, according to the spatial discretization of the field model, included 1,200 nodes situated within the detection range of the modeled microwave radiometers. This configuration resulted in a total of $5 \times 1,200 = 6,000$ hydraulic parameters being defined in the model. Although a full sensitivity analysis would typically involve all model parameters, preliminary findings showed that parameters associated with nodes beyond the specified sensing depth could not be effectively estimated using the available observations. As such, the estimability analysis was limited to parameters located within the sensing range by constructing the sensitivity matrix based only on these nodes.

The parameter vector of the field, denoted by $\mathbf{\Phi}$, was organized as:

$$\mathbf{\Phi} = [K_{s_1}, \theta_{s_1}, \theta_{r_1}, \alpha_1, n_1, \dots, K_{s_{1,200}}, \theta_{s_{1,200}}, \theta_{r_{1,200}}, \alpha_{1,200}, n_{1,200}]^T \in \mathbb{R}^{6,000}$$

Each sensitivity matrix corresponding to an azimuthal node featured 6,000 columns, each aligned with a specific model parameter. Consequently, the application SVD to these matrices produced a set of 6,000 singular values, denoted by $\sigma_1, \dots, \sigma_{6,000}$.

To assess the performance of the proposed method in estimating both soil moisture and hydraulic parameters, three test cases were developed. In Case 1, only soil moisture was estimated, with uncertainty introduced into the hydraulic parameters used in the simulation. Case 2 involved simultaneous estimation of both soil moisture and hydraulic parameters, but did not incorporate sensitivity analysis or the selection of estimable parameters. Case 3 applied the complete estimation framework, including both soil moisture and a selected subset of hydraulic parameters. The accuracy of each scenario was quantified using the root mean square error (RMSE), evaluated both at specific time steps and over the entire simulation period. The metrics used for this evaluation are defined as follows:

$$\mathrm{RMSE}_{x_a}(k) = \sqrt{\frac{\sum_{i=1}^{n_{x_a}} (\hat{x}_{a,i}(k) - x_{a,i}(k))^2}{n_{x_a}}} \tag{4.32}$$

$$\mathrm{RMSE}_{x_a} = \frac{\sum_{k=0}^{N_{\mathrm{sim}}-1} \mathrm{RMSE}_{x_a}(k)}{N_{\mathrm{sim}}} \tag{4.33}$$

The term $\mathrm{RMSE}_{x_a}(k)$, where $k = 0, \ldots, N_{\mathrm{sim}} - 1$, denotes the progression of the RMSE value at each time step throughout the simulation. In contrast, $\mathrm{RMSE}_{x_a}$ refers to the average RMSE computed over the entire simulation horizon. The variable $N_{\mathrm{sim}}$ represents the total number of time steps in the simulation period.

### 4.6.1 Simulation Results

Figure 4.3 displays the leading 900 singular values, plotted on a logarithmic scale, derived from the scaled sensitivity matrices corresponding to three azimuthal nodes in the field model. In each plot, a significant drop occurs between the 150th and 151st singular values—spanning roughly 5.8 orders of magnitude. This sharp decline implies that the 150th singular value is effectively negligible, suggesting a numerical rank of 150. Consequently, out

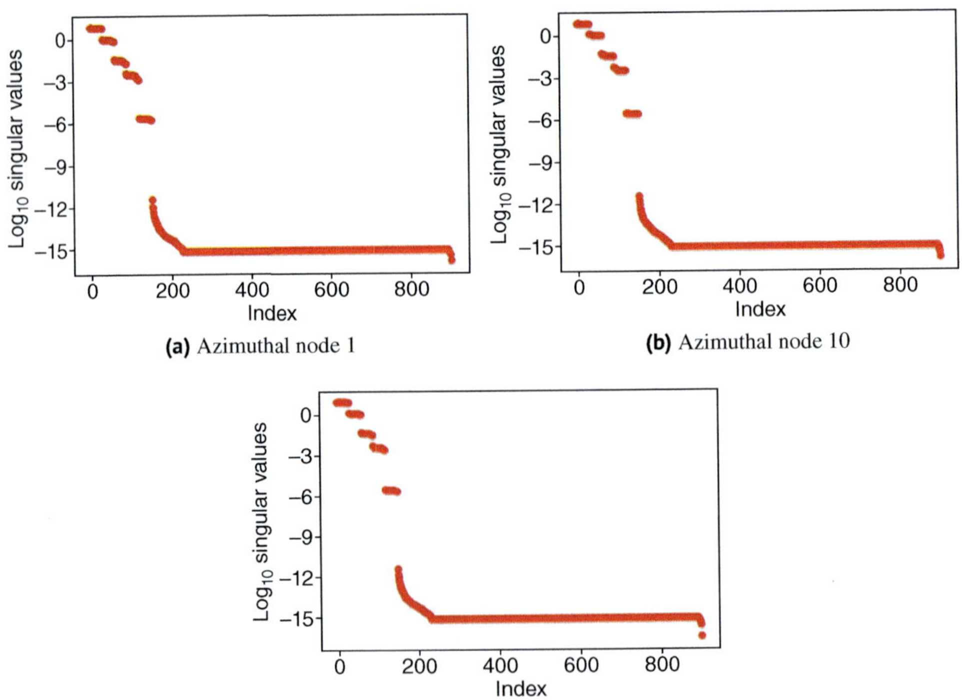


(a) Azimuthal node 1
(b) Azimuthal node 10
(c) Azimuthal node 25

**Figure 4.3** Log-scale plots of the singular values from the scaled output sensitivity matrices corresponding to selected azimuthal nodes. (a) Azimuthal node 1; (b) Azimuthal node 10; (c) Azimuthal node 25. *Source*: [79]/arXiv/CC BY 4.0.

of the 6,000 parameters considered, only 150 can be confidently identified from the available measurements at the defined sampling intervals. Similar patterns were observed in the remaining azimuthal nodes as well.

Using the orthogonal projection technique for parameter selection, five hydraulic properties ($K_s, \theta_s, \theta_r, \alpha$, and $n$) corresponding to observed field locations at each time step $k$ were identified as the most reliably estimable parameters.

Figure 4.4 summarizes the estimation results at selected field nodes over the simulation period across all three test cases. Among the cases evaluated, Case 3 demonstrated the

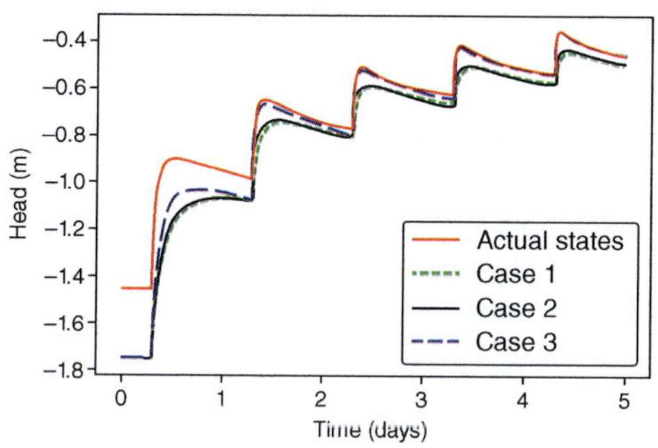


**(a)** State at node 10,700 (depth: 3.3 cm)

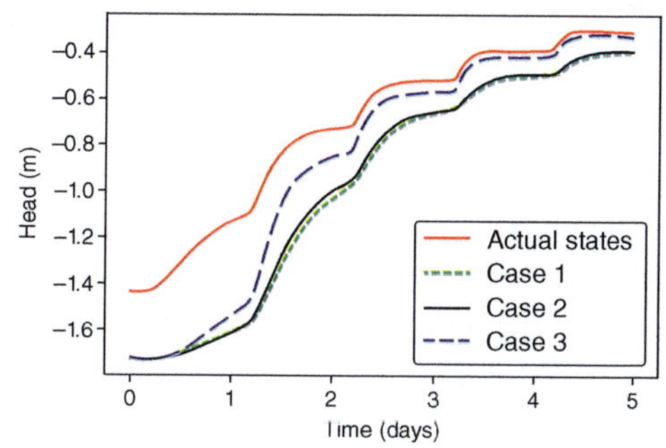


**(b)** State at node 6,601 (depth: 13.3 cm)

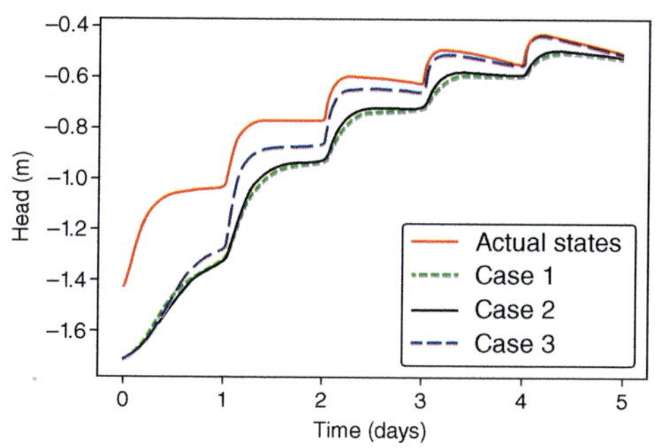


**(c)** State at node 7,200 (depth: 10 cm)

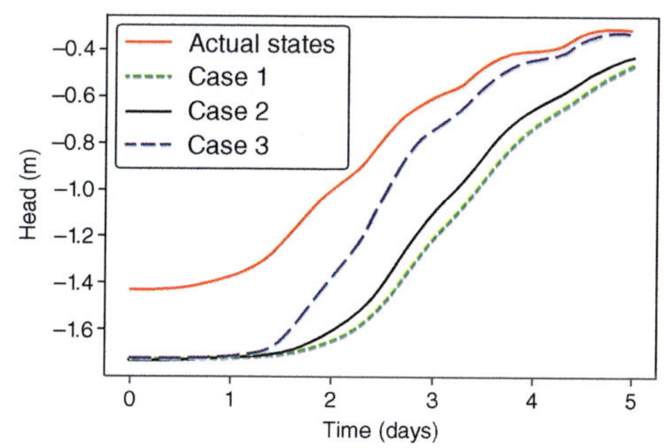


**(d)** State at node 4,150 (depth: 20 cm)

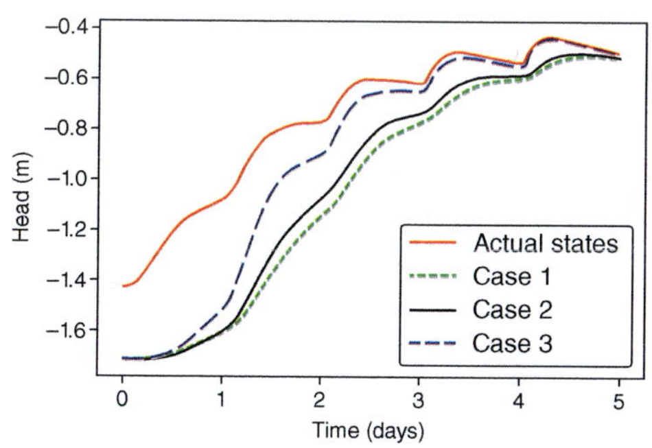


**(e)** State at node 4,840 (depth: 16.6 cm)

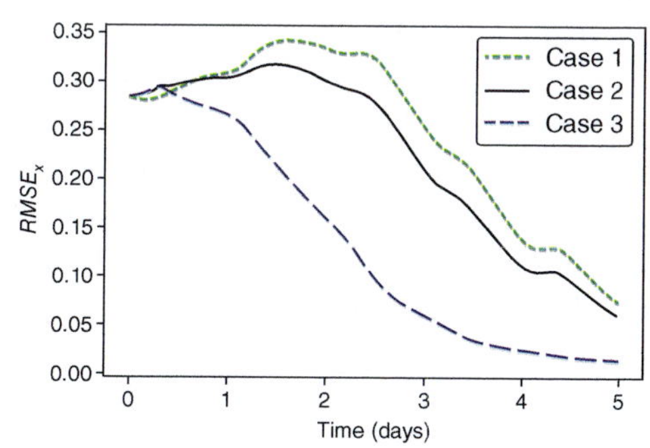


**(f)** RMSE

**Figure 4.4** (a–e) Representative trajectories showing the true states (red) alongside estimated states from Case 1 (green), Case 2 (black), and Case 3 (blue). (f) RMSE progression over time for the original state vector in Case 1 (green), Case 2 (black), and Case 3 (blue). *Source*: [79]/arXiv/CC BY 4.0.

**Table 4.1** Comparison of performance metrics across the three estimation cases.

| | $\text{RMSE}_x$ (%) | $\text{RMSE}_\theta$ (%) | $\text{RMSE}_{x_a}$ (%) |
|---|---|---|---|
| Case 1 | 26.29 | n/a | 26.29 |
| Case 2 | 24.19 | 14.20 | 24.00 |
| Case 3 | 16.60 | 13.90 | 16.44 |

highest accuracy, with estimates closely matching the true system states. Case 2 performed moderately well, while Case 1 showed the largest deviations. This ranking is supported by the RMSE values reported in Table 4.1, which indicate that the proposed method produced the most accurate soil moisture estimates.

The improvements seen in Cases 2 and 3 over Case 1 are consistent with earlier findings [34, 42], which highlight the benefit of incorporating soil hydraulic parameter estimation in improving soil moisture predictions. Furthermore, the additional gains observed in Case 3 relative to Case 2 reflect the advantages of accounting for parameter estimability. As demonstrated in previous research on joint estimation frameworks [82, 83], including parameters that are not practically estimable may degrade the accuracy of state estimates.

Although the performance of Case 3 is comparable to other advanced approaches for joint estimation of soil moisture and hydraulic properties [34, 84], its advantage lies in leveraging prior estimability analysis. As emphasized in [85], conducting such an analysis is essential to determining which parameters can be reliably estimated. By narrowing the estimation scope to only those parameters, the method avoids the inefficiencies and inaccuracies associated with poorly estimable variables, while also reducing the computational burden involved in the joint estimation process.

## 4.7 Real Case Study

This section explores the application of the proposed approach for jointly estimating system states and hydraulic parameters, using microwave remote sensing data collected from a center pivot–irrigated field. It starts by describing the physical layout of the study area and the characteristics of the sensors employed. This is followed by a detailed explanation of the numerical model used to represent the field dynamics, along with the preprocessing steps required to prepare the data for estimation. The section then discusses the sensitivity analysis performed on the model and the application of the orthogonal projection method for selecting estimable parameters. Next, the design of the estimator is introduced, along with the performance metrics used to evaluate its accuracy. The section ends with a presentation of the results and a discussion of their implications.

### 4.7.1 Study Site and Microwave Sensor Configuration

The research was carried out in 2021 at Lethbridge Polytechnic's Research Farm, situated east of Lethbridge, Alberta, Canada, at geographic coordinates 49.7230°N and 112.8001°W. The site covers an area of approximately 0.81 km$^2$ and rests at an elevation of 888 m.

The field is primarily composed of clayey loam soils, though occasional sand deposits are present. An overview of the study location is provided in Figure 4.5, with the red rectangles highlighting the specific region of interest. For modeling purposes, the analysis focused on a delineated section of Field 3, which is marked by a blue dash-dotted box in the same figure.

Soil moisture readings were obtained using a set of three passive microwave sensors, each supporting dual polarization and mounted on the center pivot system in Field 3. As illustrated in Figure 4.6, the sensors were affixed 3 m above ground level at a fixed tilt angle of 30°. The antennas featured a 65° half-power beamwidth, generating a sensing footprint approximately 4 m wide. Each unit had dimensions of 45 × 45 × 15 cm and operated within a frequency range of 380–420 MHz, enabling effective soil moisture measurements to a depth of up to 20 cm.

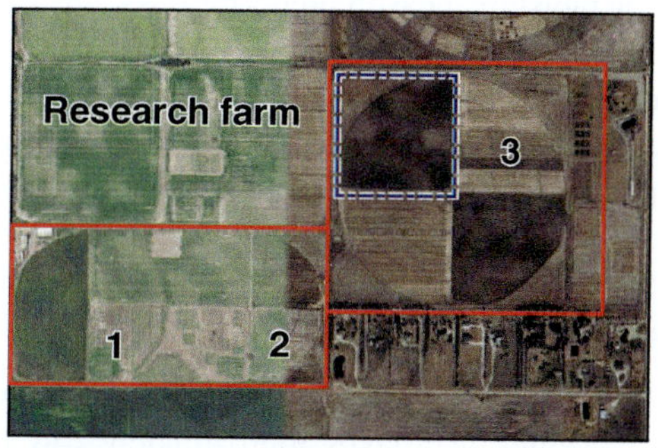


**Figure 4.5** A spatial map of the research farm managed by Lethbridge Polytechnic.
*Source*: Reproduced from [79]/arXiv/CC BY 4.0.

**Figure 4.6** Microwave radiometers installed on a center pivot irrigation system.
*Source*: Reproduced from [79]/arXiv/CC BY 4.0.

### 4.7.2 Numerical Representation of the Study Site

The section of the field examined in this study extends radially up to 290 m, extends to a vertical distance of 0.6 m, and spans an angular distance of $0.5\pi$ radians. For spatial discretization, the radial and angular dimensions were each divided into 30 and 17 equally spaced segments, respectively. The resolution in both the $r$ and $\theta$ directions was selected to align with the spatial intervals between consecutive sensor measurements. In the axial direction, the domain was discretized into 10 layers of varying thickness, with a denser grid near the surface and increasingly coarse spacing at greater depths. This configuration resulted in $N_x = 5{,}100$ state variables. The temporal dimension was discretized using a uniform time step of 10 minutes.

### 4.7.3 Sensor Data Preprocessing

Soil moisture measurements for this study were collected between June 3, 2021 and July 22, 2021, with sensors acquiring data at roughly 30-second intervals. For the purpose of jointly estimating soil moisture and hydraulic parameters, all readings obtained from June 3 to June 29 were incorporated into the data assimilation process. For the remaining simulation period, 80% of the available measurements were used for assimilation, while the remaining 20% were reserved for validation. Before integrating the data into the estimation framework, a series of preprocessing steps was applied to ensure data reliability and relevance. These steps are summarized below:

1. **Temporal Organization of Measurements**: Given the extensive area of Field 3, the center pivot takes approximately 2 to 3 days to complete a full rotation. As a result, measurements recorded near the end of a rotation spanned across multiple days. To address this, all data were initially sorted by acquisition date, followed by chronological ordering within each individual day.
2. **Quadrant-based Filtering**: Since the analysis focuses on a specific quadrant of Field 3, the dataset was refined by retaining only the measurements corresponding to that particular quadrant.
3. **Pivot Movement Reconstruction**: As the pivot system rotates, sensor positions change continuously. To approximate this movement, measurements were grouped into fixed intervals defined by $T_s$. A sampling interval of $T_s = 10$ minutes was selected to capture the spatial progression of the pivot and to enable reliable mapping of irrigated nodes at each time step.
4. **Outlier Detection and Removal**: Extreme values were identified and removed by comparing each measurement against the known saturation and residual moisture limits of the field's dominant soil type. Any data point exceeding the saturation threshold or falling below the residual moisture content was excluded.
5. **Mapping Data to Field Model Nodes**: To assign valid measurements to specific nodes in the numerical field model, a spatial grid of GPS coordinates representing the discretized model was generated. Each measurement was mapped to the closest node by calculating Euclidean distances and selecting the nearest location.

Weather-related inputs necessary for the field model—including daily reference evapotranspiration daily average temperature, and daily rainfall —were obtained from the

Alberta Information Service (see Figure A.1a–c in Appendix A.1). Irrigation amounts for the simulation period (Figure A.1c in Appendix A.1) were provided by the Research Farm. The crop coefficient values used throughout the simulation were derived using Equation (A.1), as outlined in Appendix A.4.

### 4.7.4 Analysis of Parameter Estimability and Selection Studies

For each azimuthal node within the selected quadrant, sensitivity matrices were constructed using the nominal hydraulic parameters values detailed in Figure B.1 of Appendix B.1. These matrices included a total of 2,550 parameters corresponding to nodes located within the microwave radiometer's sensing depth. The scaled sensitivity matrices exhibited rank deficiency, and the application of the orthogonal projection approach revealed that only the five hydraulic parameters linked to the observation points on the field could be uniquely estimated. These findings were used to guide the selection of estimable parameters across different time steps during the analysis of the real-world dataset.

Although the sensors originally captured data at a frequency of 30 seconds, the measurements were resampled to 10-minute intervals. Each resampled interval included between 10 and 60 observations. The orthogonal projection analysis indicated that, depending on the volume of data available at each time step, between 50 and 300 parameters could be uniquely estimated. These were computed alongside the 5,100 system states representing the discretized field model.

### 4.7.5 Estimator Design

The estimator was initialized using $\hat{x}_a(0)$ as the initial state estimate and $P_a(0|0)$ as the corresponding error covariance matrix. Due to the high degree of uncertainty associated with $\hat{x}_a(0)$, the diagonal entries of $P_a(0|0)$ were set to large values. In this configuration, the variances for the state components were assigned a value of 340, while those for the parameters were set to 6. Both the process noise covariance matrix $Q$ and the measurement noise covariance matrix $R$ were considered tuning variables within the estimation framework. These matrices were iteratively adjusted by modifying their diagonal entries to ensure that the soil moisture estimates closely aligned with observed data. Specifically, $Q$ was defined as $10I_{7,650}$, where $I_{7,650}$ denotes the identity matrix of dimension 7,650, and $R$ was specified as $0.01I_{N_y}$, with $N_y$ indicating the number of observations collected at each sampling interval.

### 4.7.6 Evaluation Criteria

The effectiveness of the proposed estimation method was tested using two cross-validation strategies. In the first, soil moisture observations at each sampling time were randomly divided into two groups: one for training and the other for validation. The estimator was applied to the training data, and the resulting soil moisture estimates were compared against the validation data to assess accuracy.

In the second approach, all observations collected on July 21, 2021, were used for validation. The model was run using actual irrigation and weather conditions from that day. Predicted soil moisture values were compared with the sensor readings collected on the same date. The absolute differences between predicted and observed values were calculated to evaluate the model's accuracy after convergence.

To measure performance, the normalized root mean square error (NRMSE) was used. This metric compared the estimated soil moisture $\hat{y}_k$ with the observed values $y_k$ at each time step and was computed using the following formula:

$$\text{NRMSE} = \frac{1}{(y_{\max} - y_{\min})} \sqrt{\frac{\sum_{k=1}^{k=N} (y_k - \hat{y}_k)^2}{N}} \tag{4.34}$$

where $N$ represents the total number of measurements in the validation set, and $y_{\max}$ and $y_{\min}$ are the maximum and minimum soil moisture values within the validation set, respectively.

To further highlight the effectiveness of the proposed method, its results are compared with those from two alternative case studies. In Case Study 1, soil moisture was estimated using fixed hydraulic parameters representative of the field's dominant soil type (sandy clay loam). Table 4.2 summarizes the hydraulic parameters of sandy clay loam. Case Study 2 used spatially varying hydraulic parameters obtained from a soil texture survey carried out in the field. The distribution of these parameters is shown in Figure B.2 of Appendix B.2.

### 4.7.7 Experimental Results and Discussion

#### 4.7.7.1 Estimation Accuracy

Figures 4.7 and 4.8 present the outcomes of the first cross-validation method conducted on July 2 and July 5. Each plot includes the reference line ($y = x$), allowing for direct visual comparison across the three case studies. A close inspection of the figures, alongside the corresponding NRMSE values, indicates that the proposed method achieves the highest level of accuracy, followed by Case 2 and then Case 1. This ranking is evident from the distribution of the validation points: in Case 3, the points are closely clustered around the $y = x$ line, whereas Cases 2 and 1 show increasingly scattered patterns. When using Case 1 as a baseline, the proposed method improves estimation accuracy by 43% on July 2 and by 24% on July 5.

The lower accuracy observed in Case 1 highlights the limitations of relying on fixed hydraulic parameters associated with the dominant soil type, as discussed in [81]. A single parameter set is insufficient to capture the spatial heterogeneity of soil properties across the field. The improved performance in Cases 2 and 3 supports the conclusions of [34], which emphasized that errors in hydraulic parameter values can introduce bias and degrade estimation quality. The additional performance gap between Case 2 and Case 3 is consistent

**Table 4.2** Hydraulic properties of sandy clay loam.

| $K_s$ (m/s) | $\theta_s$ (m³/m³) | $\theta_r$ (m³/m³) | $\alpha$ (1/m) | $n$ (-) |
|---|---|---|---|---|
| $7.222 \times 10^{-7}$ | 0.410 | 0.090 | 1.90 | 1.31 |

*Source:* [86].

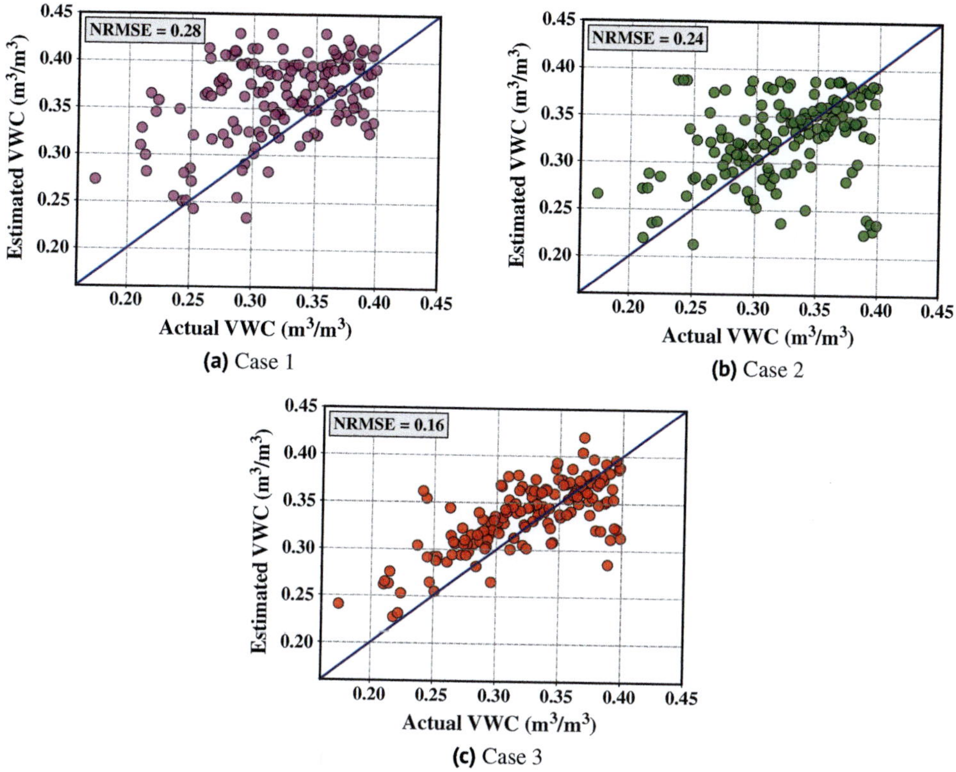


**Figure 4.7** Results of the cross-validation conducted on July 2, 2021. (a) Case 1; (b) Case 2; (c) Case 3. *Source*: [79]/arXiv/CC BY 4.0.

with findings in [87, 88], which point to the shortcomings of pedotransfer functions when used alone.

#### 4.7.7.2 Predictive Accuracy

In the second cross-validation, Case Study 1 was excluded due to its poor estimation performance in the first cross-validation. The comparison focused on Case Study 2 and the proposed method, and the validation outcomes are illustrated in Figures 4.9 and 4.10, and corresponding NRMSE values reported in Table 4.3. As shown in Figures 4.9c and 4.10c, the proposed method demonstrated significantly better accuracy, with a maximum absolute error of approximately 0.086 $m^3/m^3$, compared to 0.20 $m^3/m^3$ for Case Study 2. This improvement is also reflected in the NRMSE values, where the proposed approach achieved the lowest error. Compared to Case Study 2, the predictive accuracy improved by 50%.

These results support the findings of [81], which emphasize the usefulness of remote sensing for estimating near-surface soil moisture and providing reliable short-term prediction models. This is particularly beneficial in settings where direct field measurements are unavailable but accurate soil moisture data are still needed for irrigation decision-making.

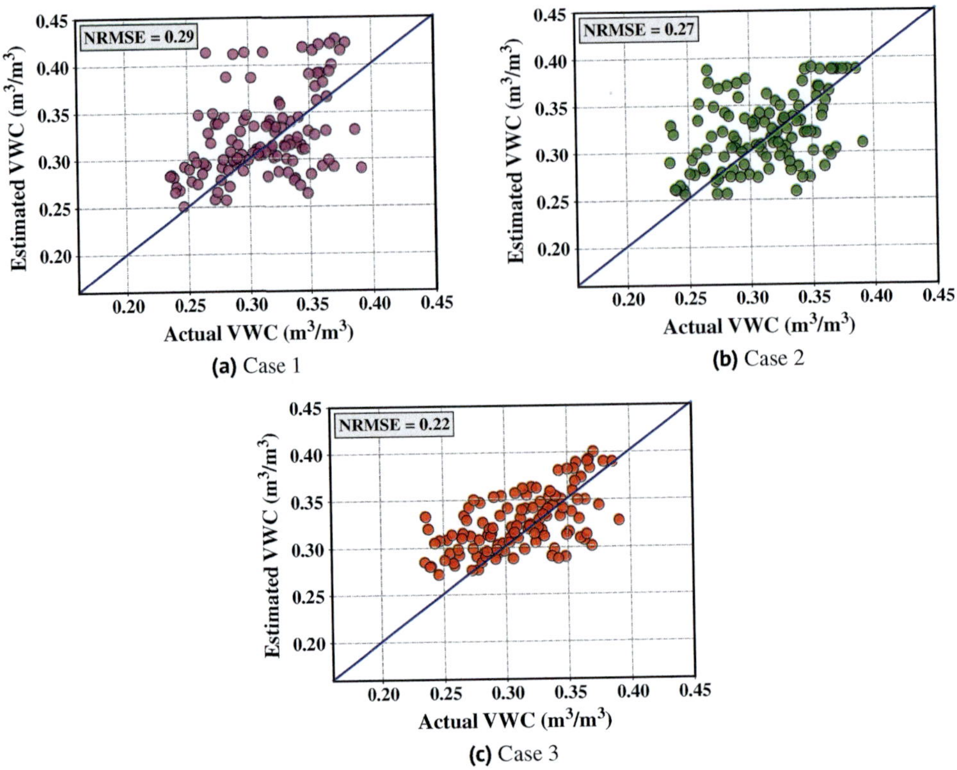


**Figure 4.8** Results of the cross-validation conducted on July 5, 2021. (a) Case 1; (b) Case 2; (c) Case 3. *Source*: [79]/arXiv/CC BY 4.0.

#### 4.7.7.3 Accuracy of Estimated Parameters

Figures 4.11 and 4.12 display the spatial distribution of two key hydraulic parameters: saturated hydraulic conductivity ($K_S$) and saturated volumetric moisture content. The study placed particular emphasis on evaluating the accuracy of $K_s$ estimates. According to Figure 4.11, the estimated values of $K_s$ ranged from 0.261 to 0.336 m/day. To assess the reliability of these estimates, laboratory measurements were obtained from five randomly selected locations and compared with the estimated values. The mean laboratory measurement for $K_s$ was 0.39 m/day, showing good agreement with the estimated values and suggesting that the proposed method provides representative estimates for the broader field area.

Although Figure 4.11 illustrates the framework's effectiveness in capturing horizontal variability in soil hydraulic properties, it also exposes a key limitation: the estimation approach does not resolve vertical variation. This limitation stems from the nature of the observations employed for the estimation process. The microwave radiometers employed in this study provide soil moisture measurements that represent an average over the top 20 cm of the soil profile. Since only one measurement is available per location in the vertical direction, and the parameter selection process links estimable parameters to the observation locations, the estimates reflect depth-averaged conditions. Consequently, the current framework cannot capture vertical gradients in hydraulic parameters. This limitation highlights the need for complementary sensing approaches. Incorporating point-based

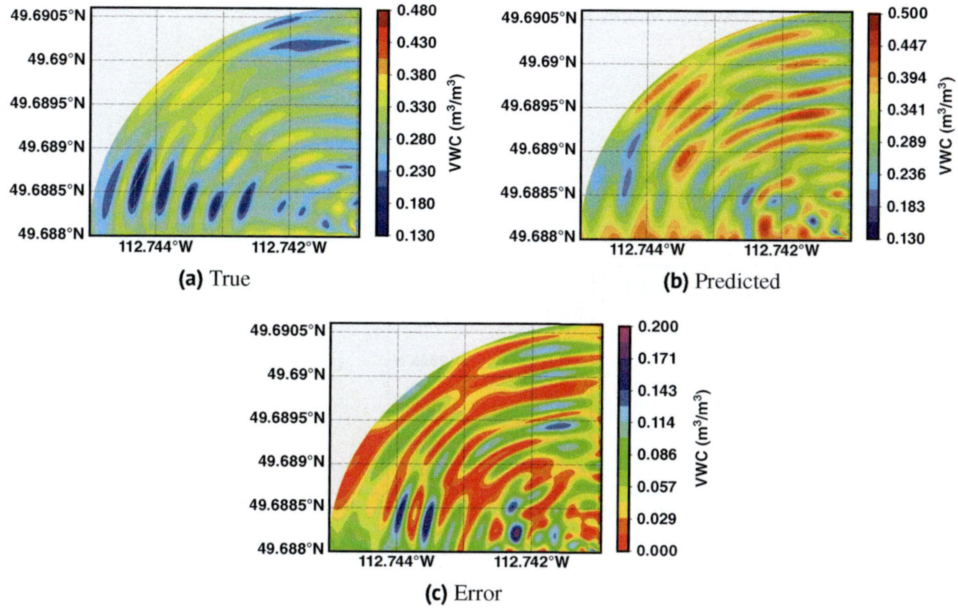


**Figure 4.9** Case 2 cross-validation with July 21, 2021 measurements. (a) True; (b) Predicted; (c) Error. *Source*: [79]/arXiv/CC BY 4.0.

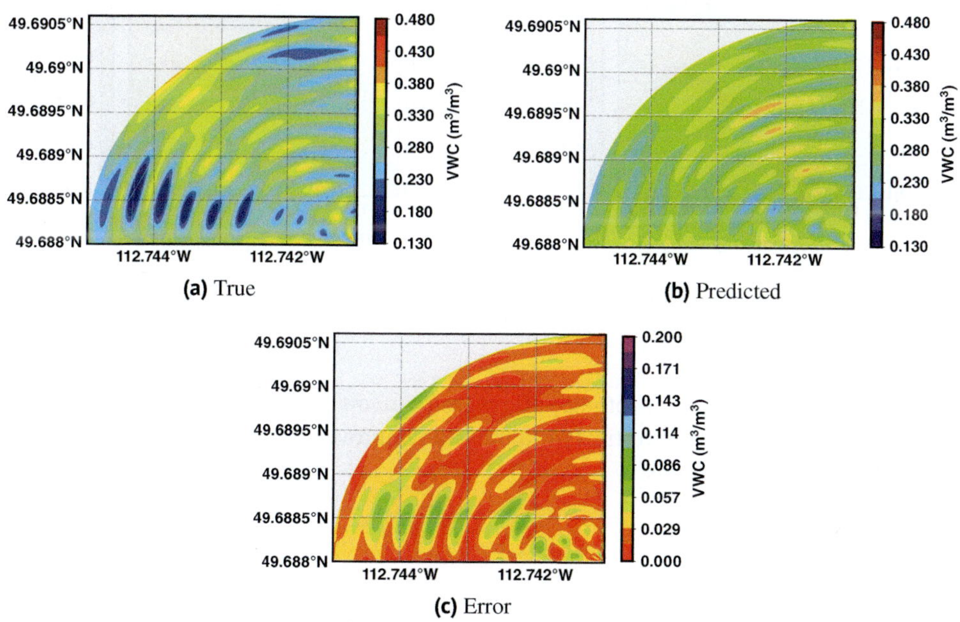


**Figure 4.10** Case 3 cross-validation with July 21, 2021 measurements. (a) True; (b) Predicted; (c) Error. *Source*: [79]/arXiv/CC BY 4.0.

**Table 4.3** Summary of NRMSE values from July 21, 2021 Type 2 validation.

| Scenario | NRMSE |
|---|---|
| Estimation with texture survey parameters (Case 2) | 0.24 |
| Proposed approach (Case 3) | 0.12 |

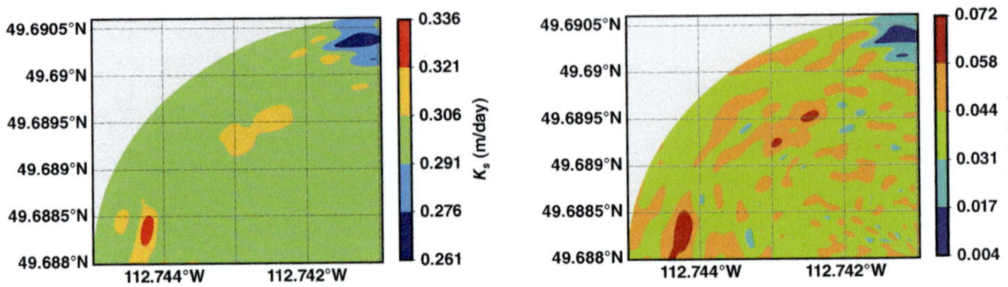


**(a)** Spatial map of $\hat{K}_s$. **(b)** Uncertainty map of $\hat{K}_s$

**Figure 4.11** Estimated $K_s$ (m/day) across the field at the end of the simulation. (a) Spatial map of $\hat{K}_s$; (b) Uncertainty map of $\hat{K}_s$. *Source:* [79]/arXiv/CC BY 4.0.

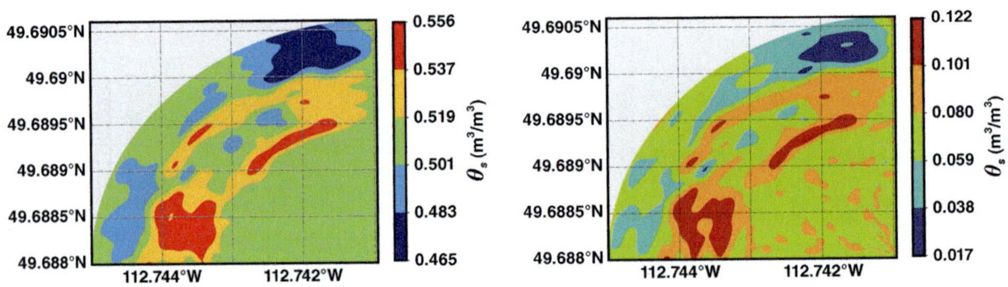


**(a)** Spatial map of $\hat{\theta}_s$ **(b)** Uncertainty map of $\hat{\theta}_s$

**Figure 4.12** Estimated $\theta_s$ (m$^3$/m$^3$) across the field at the end of the simulation. (a) Spatial map of $\hat{\theta}_s$; (b) Uncertainty map of $\hat{\theta}_s$. *Source*: [79]/arXiv/CC BY 4.0.

sensors capable of capturing depth-specific measurements could enhance the estimation framework's ability to resolve vertical variability and improve overall model performance in agro-hydrological applications.

## 4.8 Conclusion

This chapter outlines a framework for estimating soil moisture and hydraulic parameters across large-scale agricultural fields. The methodology relied on remotely sensed soil moisture data obtained from microwave radiometers mounted on center pivot systems. To represent moisture movement in the soil, the Richards' equation was adapted for a cylindrical coordinate system. A modified sensitivity analysis was used to evaluate issues related to the estimability of hydraulic parameters, and orthogonal projection was applied to select a subset of parameters suitable for assimilation. Using an EKF, both soil moisture states and the selected hydraulic parameters were estimated by assimilating field measurements into the model. The approach was tested in both simulated environments and real field conditions, demonstrating improvements in soil moisture estimation and producing consistent parameter values. The results presented in this chapter support practical applications such as irrigation management zone delineation and model calibration, both of which are essential for developing effective irrigation scheduling systems.

# 5

# Adaptive Soil Moisture Estimation Using Performance-triggered Model Reduction

## 5.1 Introduction

Accurately estimating soil moisture remains essential for designing efficient closed-loop irrigation systems. In Chapter 4, a soil moisture estimation approach was introduced that assimilates field observations into a cylindrical-coordinate form of the Richards' equation. This enabled the inference of spatially distributed soil moisture profiles alongside a select group of hydraulic parameters. Despite the effectiveness of this estimation approach, several challenges, particularly in relation to computational complexity of the scheme, persist.

The Richards' equation, which has found widespread application in the numerical simulation of soil moisture dynamics in agro-hydrological systems, is inherently difficult to solve. Its highly nonlinear nature, coupled with its reliance on complex constitutive relations, places a heavy computational burden on any estimation scheme that relies on it. Moreover, ensuring numerical stability and precision often requires fine spatial discretization, especially near the soil surface, which further increases model complexity. These factors collectively limit the scalability and practical deployment of Richards' equation-based estimation schemes in large agricultural systems.

Amid these constraints, reducing model complexity while preserving estimation accuracy has become a key priority. Traditional model reduction techniques such as proper orthogonal decomposition [89], sensitivity-based reduction [90], Hankel-norm-based approximations [91, 92], and balanced truncation [93] offer various ways of simplifying dynamical systems. However, these methods often yield reduced models whose internal states no longer correspond to physically meaningful quantities. This limitation is particularly relevant in agro-hydrological settings, where the interpretability of state variables holds practical importance.

Structure-preserving approaches provide a more interpretable alternative. Among these, clustering-based methods [94] have proven effective in maintaining both the physical meaning of states and the topological structure of the model. These techniques have recently been applied to agro-hydrological models [95, 96], where reduced-order representations derived from Richards'-based simulations have been embedded within moving horizon estimators. While these efforts mark a significant step forward, they rely on offline model construction under predefined simulation scenarios. Once deployed, such models may underperform if field conditions differ from those used during development. Although periodic updates

*Precision Irrigation for Agriculture: Integrating Machine Learning and Optimal Control Strategies*,
First Edition. Bernard Twum Agyeman and Jinfeng Liu.

using new trajectory data can improve consistency, the added computational load can be substantial.

This chapter introduces a more flexible alternative: a performance-triggered adaptive model reduction approach for soil moisture estimation. Rather than depending on a fixed reduced model, the framework continuously monitors prediction accuracy and updates the model only when performance drops below a specified threshold. The full Richards' equation remains in use as a high-fidelity simulator, while unsupervised learning is applied to cluster state trajectories with similar dynamic behavior. When the reduced model's predictions diverge beyond a defined threshold, a new model is generated to restore accuracy. Estimation across the field is then carried out using an adaptive reduced-order extended Kalman filter (EKF). The chapter concludes with a case study involving a large-scale agricultural field under varying conditions. The simulation results, based in part on earlier work reported in [97], highlight the effectiveness of the proposed adaptive framework in maintaining both efficiency and accuracy over time.

## 5.2 Model Development and Problem Statement

This chapter focuses on an agro-hydrological system equipped with a center pivot irrigation system, as illustrated previously in Figure 4.1 of Chapter 4. Accordingly, the field model adopted here is the three-dimensional Richards' equation expressed in cylindrical coordinates, originally presented in Chapter 2. For completeness, and to support a clearer understanding of the discussions that follow, the cylindrical-coordinate formulation of the Richards' equation is restated below:

$$\begin{aligned} C(\psi)\frac{\partial \psi}{\partial t} &= \frac{1}{r}\frac{\partial}{\partial r}\left[rK(\psi)\frac{\partial \psi}{\partial r}\right] + \frac{1}{r}\frac{\partial}{\partial \theta}\left[\frac{K(\psi)}{r}\frac{\partial \psi}{\partial \theta}\right] + \\ &\quad \frac{\partial}{\partial z}\left[K(\psi)\left(\frac{\partial \psi}{\partial z}+1\right)\right] - S(\psi, z) \end{aligned} \tag{5.1}$$

Although a comprehensive derivation and explanation of Equation (5.1) are presented in Chapter 2, several key terms are restated here to support clearer appreciation of the material that follows. In particular, $\psi$ denotes the pressure head (m); $r$, $\theta$, and $z$ represent the radial, azimuthal, and axial spatial coordinates, respectively; and $t$ denotes the temporal variable.

Equation (5.1) is numerically solved using a central difference scheme to approximate spatial derivatives along the $r$, $\theta$, and $z$ directions. A detailed explanation of this numerical approach is provided in Section 4.2 of Chapter 4, which outlines the solution procedure for the cylindrical-coordinate form of the Richards' equation. As such, this chapter does not revisit the full numerical formulation, and the reader is directed to the earlier section for further details. The boundary conditions employed in the numerical simulation are given below:

$$\frac{\partial \psi(r,\theta,z)}{\partial r} = 0 \qquad \text{at } (r=0,\ \theta,\ z) \tag{5.2}$$

$$\frac{\partial \psi(r,\theta,z)}{\partial r} = 0 \qquad \text{at } (r=H_r,\ \theta,\ z) \tag{5.3}$$

$$\frac{\partial (\psi(r,\theta,z)+z)}{\partial z} = 1 \qquad \text{at } (r,\ \theta,\ z=0) \tag{5.4}$$

$$\frac{\partial(\psi(r,\theta,z))}{\partial z} = -1 - \frac{u^{\mathrm{irr}} - \mathrm{EV}}{K(\psi)} \quad \text{at } (r,\ \theta,\ z = H_z) \tag{5.5}$$

$$\psi(r,\ \theta = 0,\ z,\ t) = \psi(r,\ \theta = 2\pi,\ z,\ t) \tag{5.6}$$

In Equation (5.3), $H_r$ denotes the total radius of the irrigated field, whereas $H_z$ in Equation (5.5) specifies the vertical extent of the soil profile. The terms $u^{\mathrm{irr}}$ (m/s) and EV (m/s), appearing in Equation (5.5), represent the irrigation amount and the potential evapotranspiration, respectively.

After numerical discretization, the field model is expressed in state-space form as follows:

$$\dot{x}(t) = f(x(t), u(t), \Phi) + \omega(t) \tag{5.7}$$

$$y(t) = Cx(t) + v(t) \tag{5.8}$$

In Equation (5.7), $x(t) \in \mathbb{R}^{N_x}$ denotes the state vector of the system, consisting of $N_x$ pressure head values. The input vector $u(t) \in \mathbb{R}^{N_u}$ and the model disturbance $\omega(t) \in \mathbb{R}^{N_x}$ represent the external forcing (such as irrigation, rainfall, reference evapotranspiration, and crop coefficient) and unmodeled dynamics, respectively. The vector of hydraulic parameters is denoted by $\Phi \in \mathbb{R}^{N_p}$, and includes the saturated hydraulic conductivity ($K_s$), residual moisture content ($\theta_r$), saturated moisture content ($\theta_s$), and the shape parameters ($\alpha, n$).

The estimation framework assumes that direct measurements of the pressure head are available. As described in Equation (5.8), $y(t) \in \mathbb{R}^{N_y}$ denotes the vector of pressure head measurements at time $t$, $C \in \mathbb{R}^{N_y \times N_x}$ is the output matrix mapping the system states to measurements, and $v(t) \in \mathbb{R}^{N_y}$ represents measurement noise. In practice, it is reasonable to consider direct pressure head measurements, as tensiometers can be deployed in the field to obtain such data. Typically, the number of available measurements is significantly smaller than the number of states, i.e., $N_y \ll N_x$, and the goal of the estimation task is to reconstruct the complete soil moisture state from this limited set of observations.

As previously noted, the Richards' equation poses considerable computational challenges due to its nonlinear structure and high dimensionality. These properties hinder its direct use within soil moisture estimation frameworks, especially when applied to large-scale agricultural systems. A practical way forward involves simplifying the field model through model reduction techniques that reduce both dimensionality and computational effort before developing the estimation scheme.

From a practical application point of view, an effective model reduction approach must satisfy two essential criteria. First, it should preserve the system's topological structure and retain the physical interpretability of the state variables, making structure-preserving techniques a suitable choice. Second, the method must reflect the variability in soil moisture dynamics observed throughout the growing season. Agricultural fields are subject to a wide range of environmental and operational conditions, which influence soil moisture in complex ways over time. A single reduced model generated offline is unlikely to remain valid across such variability, as the offline trajectories used in its construction may not capture all relevant seasonal conditions.

In response to these challenges, this chapter adopts a performance-triggered adaptive model reduction strategy based on a structure-preserving framework. The resulting reduced-order model offers a computationally efficient representation of the field dynamics while maintaining adaptability to evolving field conditions. This model is subsequently used to estimate soil moisture in a large-scale agricultural setting.

## 5.3 Model Reduction and Estimation Scheme

This section provides a detailed exposition of the performance-triggered adaptive model reduction strategy and its integration with an EKF framework for soil moisture estimation. The overall procedure, including the key stages of model reduction and estimation, is outlined in Figure 5.1. The proposed algorithm comprises four principal components, which are summarized below:

1. Simulation of state trajectories using the original high-fidelity model,
2. Construction of a reduced-order model to approximate the system dynamics,
3. Evaluation of a performance-based criterion for triggering model reduction, and
4. Design of an EKF based on the current reduced model to estimate the system state $\hat{\tilde{x}}$.

In this framework, the model error $e_L$ serves as a metric for assessing the predictive accuracy of the reduced-order model. When the reduced model fails to adequately capture the evolving soil moisture dynamics, often due to changes in environmental or irrigation conditions, $e_L$ increases. Once $e_L$ exceeds a predefined threshold $th_e$, the system uses available inputs, including scheduled irrigation and weather forecasts, to regenerate state trajectories through the original model. These updated trajectories are then used to construct a new reduced-order model that better reflects current conditions. This trajectory-driven update mechanism enhances the model's ability to track variations in field conditions and improves overall estimation accuracy. Soil moisture estimation across the field is subsequently performed using the updated reduced-order model in conjunction with field measurements. The remainder of this section elaborates on each component of the proposed framework in detail.

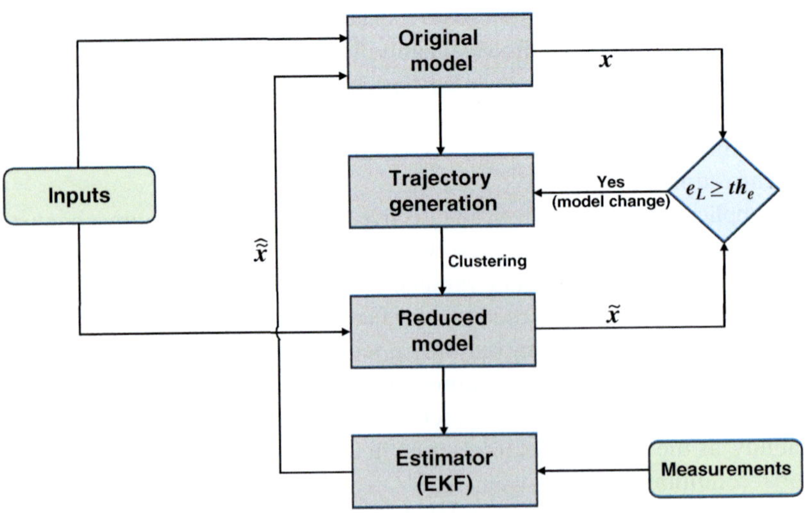


**Figure 5.1** Overview of the performance-triggered adaptive model reduction and state estimation framework. *Source*: [97]/John Wiley & Sons/CC BY 4.0.

### 5.3.1 Adaptive Structure-preserving Model Reduction

When the value of $e_L$ exceeds the threshold $th_e$ at a particular time step $t_k$, the reduced model is regenerated. The steps involved in this process are identical to those used during the initial model construction and are presented below:

#### 5.3.1.1 Step 1: Generation of State Trajectories

Reduced model construction begins by generating a sequence of state trajectories across a designated time horizon comprising $N_{f_d}$ discrete steps. This is done by simulating the system dynamics described in Equation (5.7), using as inputs the planned irrigation schedule and forecasted weather conditions over the time window. The simulation is initialized using the most recent estimate of the system state at time $t_k$. In practical settings, irrigation inputs are typically derived from model-based irrigation scheduling schemes, while weather forecasts provide the necessary environmental data. The resulting sequence of system states is represented by $X_m$, defined as follows:

$$X_m \ := \begin{bmatrix} x(t_k) & x(t_{k+1}) & \cdots & x(t_{k+N_{f_d}}) \end{bmatrix}^\top$$

The matrix $X_m \in \mathbb{R}^{(N_{f_d}+1)\times N_x}$ contains the collected state snapshots corresponding to the $m$th model reduction, assuming that $m-1$ reductions have already occurred prior to the current time step $t_k$. Each reduced model obtained through this process is expected to provide reliable approximations of system behavior for at least $N_{f_d}$ future sampling intervals.

#### 5.3.1.2 Step 2: Cluster-based Model Reduction

A new reduced-order model is constructed using the snapshot matrix $X_m$. The core idea behind the clustering step is to group state trajectories exhibiting similar behavior, thereby allowing each cluster, rather than individual state variables, to represent a single state in the reduced model. An agglomerative hierarchical clustering is employed to perform this grouping, following the methodology outlined in [98]. The procedure begins by treating each state trajectory as an independent cluster. Pairwise distances between clusters are then computed, and clusters separated by a distance smaller than a predefined threshold $\delta_c$ are iteratively merged into a single cluster. The threshold $\delta_c$ serves as a key tuning parameter, as it directly affects the resolution and predictive performance of the resulting reduced-order model. For evaluating the similarity between trajectories, Euclidean distance is used as the metric.

Assume that clustering results in $r_m$ clusters following the $m$th model reduction. Let the cluster set be denoted by $C^{(m)} = \{C_1^{(m)}, C_2^{(m)}, \dots, C_{r_m}^{(m)}\}$. The construction of the $m$th reduced-order model is based on the Petrov–Galerkin projection method, as described in [92]. Central to this approach is the projection matrix $U^{(m)} \in \mathbb{R}^{N_x \times r_m}$, whose structure is directly informed by the cluster configuration $C^{(m)}$. The entries of $U^{(m)}$ are defined as:

$$U_{i,j}^{(m)} = \begin{cases} w_i, & \text{if } i \in C_j^{(m)} \\ 0, & \text{otherwise} \end{cases}$$

Here, $w_i$ is a weighting factor associated with the $i$th state variable in cluster $C_j^{(m)}$. It is computed as:

$$w_i = \frac{1}{||\alpha_i||}, \quad \alpha_i = \mathbb{E}_i^\top \alpha$$

where $\alpha = [1, \ldots, 1]^\top \in \mathbb{R}^{N_x}$ is an inclusion vector, and $||\alpha_i||$ denotes the $L_2$ norm of $\alpha_i$. The matrix $\mathbb{E}_i = e_{C_i} \in \mathbb{R}^{N_x \times N_{C_i}}$ contains the standard basis vectors corresponding to the indices in cluster $C_i$, with each $e_j$ being a column of the identity matrix $I_{N_x}$.

Consequently, the reduced dynamics are formulated as:

$$\dot{\xi}^{(m)}(t) = f_r^{(m)}(\xi^{(m)}(t), u(t)) + w_r(t) \tag{5.9}$$

where $f_r^{(m)} := U^{(m)^\top} f$ denotes the reduced nonlinear function, $\xi^{(m)}(t) := U^{(m)^\top} x(t)$ represents the state vector of the reduced-order model, obtained by projecting the full-order state onto a lower-dimensional subspace and $w_r(t)$ is the process uncertainty in the reduced state.

By applying numerical discretization, the reduced model is expressed in discrete-time form as:

$$\xi_{k+1}^{(m)} = f_{rd}^{(m)}(\xi_k^{(m)}, u_k) + w_{r_k} \tag{5.10}$$

where $f_{rd}^{(m)}$ denotes the discrete-time reduced dynamics. The full-state $\tilde{x}(t)$ can then be approximated from the reduced state using the projection matrix, as follows:

$$\tilde{x}(t) := U^{(m)} \xi^{(m)}(t)$$

When transitioning between reduced-order models, the dimensionality of the reduced state space may differ. A two-step procedure is applied to maintain consistency during this transition. First, the current reduced-state estimate is mapped back into the full-order state space. Next, the resulting full-state estimate is projected onto the new reduced-order subspace using the most recent projection matrix.

### 5.3.2 Reduced-order Adaptive EKF

As previously discussed, the dimension of the reduced-order model may vary following model updates or re-identification. This change prevents the direct application of the standard EKF algorithm. In response, an adaptive EKF is developed that operates in conjunction with the time-varying reduced-order model. The discrete-time reduced model in Equation (5.10) is therefore combined with an output equation, defined as follows:

$$\begin{aligned} \xi_{k+1}^{(m)} &= f_{rd}^{(m)}\left(\xi_k^{(m)}, u_k\right) + w_{r_k} \\ y_k &= C_r^{(m)} \xi_k^{(m)} + v_k \end{aligned} \tag{5.11}$$

where $C_r^{(m)} := C U^{(m)}$.

At sampling time $t_k$, the prediction step of the adaptive EKF begins by computing a forecast of the reduced-order state $\xi_{k|k-1}^{(m)}$, using the state estimate from $t_{k-1}$, i.e., $\xi_{k-1|k-1}^{(m)}$ and the dynamics of the reduced model, as given below:

$$\hat{\xi}_{k|k-1}^{(m)} = f_{rd}^{(m)}\left(\hat{\xi}_{k-1|k-1}^{(m)}, u_k\right) \tag{5.12}$$

The covariance of the reduced-order state estimate is likewise propagated using the reduced model dynamics:

$$P_r^{(m)}(k|k-1) = A_{k-1}^{(m)} P_r^{(m)}(k-1|k-1) A_{k-1}^{(m)\top} + Q_r^{(m)} \tag{5.13}$$

In this context, $P_r^{(m)}$ and $Q_r^{(m)}$ represent the state estimation covariance and process noise covariance matrices, respectively, within the reduced-order space. The matrix $A_{k-1}^{(m)} := \frac{\partial f_{rd}^{(m)}}{\partial \xi^{(m)}}\Big|_{\hat{\xi}_{k-1|k-1}^{(m)}, u_k}$ corresponds to the Jacobian of the discrete-time reduced dynamics, evaluated at $\hat{\xi}_{k-1|k-1}^{(m)}$ and $u_k$. The initialization of the adaptive EKF relies on projecting the covariance information from the full-order system into the reduced space. Let $P(0|0)$ and $Q$ denote the initial state covariance and process noise covariance matrices associated with the full-order model. Their reduced-order counterparts are defined as $P_r^{(1)}(0|0) := {U^{(1)}}^\top P(0|0) U^{(1)}$ and $Q_r^{(m)} := {U^{(m)}}^\top Q U^{(m)}$, respectively.

In the update step of the EKF, the current reduced-state estimate $\hat{\xi}_{k|k}^{(m)}$ is computed using $\hat{\xi}_{k|k-1}^{(m)}$ and the available measurement $y_k$, as given by:

$$\hat{\xi}_{k|k}^{(m)} = \hat{\xi}_{k|k-1}^{(m)} + K_{r_k}^{(m)} \left( y_k - C_r^{(m)} \hat{\xi}_{k|k-1}^{(m)} \right) \tag{5.14}$$

where $K_{r_k}^{(m)}$ is the Kalman gain associated with the $m$th reduced-order model, and is computed as:

$$K_{r_k}^{(m)} = P_r^{(m)}(k|k-1) C_r^{(m)\top} \left( R + C_r^{(m)} P_r^{(m)}(k|k-1) C_r^{(m)\top} \right)^{-1} \tag{5.15}$$

where $R$ denotes the covariance matrix of the measurement noise. The state estimation covariance is then updated according to:

$$P_r^{(m)}(k|k) = \left( I_{r_m} - K_{r_k}^{(m)} C_r^{(m)} \right) P_r^{(m)}(k|k-1) \tag{5.16}$$

where $I_{r_m}$ is the identity matrix of size $r_m$. The matrices $P(0|0)$, $Q$, and $R$ serve as tuning parameters for the EKF algorithm. Finally, the full-state estimate at time $t_k$, denoted by $\hat{\hat{x}}_{k|k}$, is obtained by projecting the reduced-state estimate back into the full-order space:

$$\hat{\hat{x}}_{k|k} = U^{(m)} \hat{\xi}_{k|k}^{(m)} \tag{5.17}$$

When the reduced-order model is updated, the internal states and covariance estimates maintained by the EKF must be transferred consistently from the previous model to the newly constructed one. This process involves two main steps. First, the reduced-order state estimate and its covariance are mapped back to the full-order state space. Then, this full-order representation is projected onto the subspace defined by the new reduced-order model. Assuming the EKF is transitioning from the $m$th to the $(m+1)$th reduced model, the following operations are performed:

1. **Reconstruction in Full-order Space:**

$$\hat{\hat{x}} = U^{(m)} \hat{\xi}^{(m)}, \quad P = U^{(m)} P_r^{(m)} {U^{(m)}}^\top$$

2. **Projection into New Reduced Space:**

$$\hat{\xi}^{(m+1)} = {U^{(m+1)}}^\top \hat{\hat{x}}, \quad P_r^{(m+1)} = {U^{(m+1)}}^\top P U^{(m+1)}$$

These values are used to initialize the EKF corresponding to the $(m+1)$th reduced-order model.

### 5.3.3 Error Metric and Implementation Algorithm

The update of the reduced model is governed by an error metric $e_L$, which is evaluated at each sampling time $t_k$ to assess the adequacy of the current reduced model. Specifically, the estimated full-state vector $\hat{x}(t_k)$ serves as the initial condition for simulating both the original model and the reduced-order model over a prediction horizon of $N_{fd}$ steps. An open-loop irrigation profile, assumed to be known in advance, is used in these simulations.

The deviation between the trajectories produced by the two models is then computed to quantify the performance of the reduced model. The error metric is defined as:

$$e_L(t_k) = \frac{1}{N_x} \sum_{j=1}^{N_{fd}} \sum_{i=1}^{N_x} |\tilde{x}_i(t_{k+j}) - x_i(t_{k+j})| \tag{5.18}$$

where $\tilde{x}_i$ and $x_i$ denote the $i$th state predictions from the reduced and original models, respectively.

A new reduced model is constructed whenever $e_L(t_k)$ exceeds a threshold $th_e$. In addition to $e_L$, a secondary metric $\Delta e_L$ is introduced to account for the recent trend in error values and to limit sensitivity to disturbances. This metric is computed as:

$$\Delta e_L(t_k) = \frac{1}{N_e} \sum_{j=0}^{N_e-1} (e_L(t_{k-j}) - e_L(t_{k-j-1}))$$

with $N_e = 10$ in this chapter.

Algorithm 1 summarizes the key steps that make up the performance-triggered reduced soil moisture estimator.

---

**Algorithm 1** Performance-triggered Reduced EKF Algorithm

---

**Require:** Initial estimate $\hat{x}_{0|0}$, $P(0|0) > 0$, noise covariances $Q, R > 0$, threshold $th_e$, clustering threshold $\delta_c$, $m = 0$, and prediction horizon $N_{f_d}$

```
for k = 0...n do
    if e_L(k) > th_e ∧ Δe_L(k) ≥ 0.05 ∨ k = 0 then
        m ← m + 1
        Generate snapshot matrix X_m using x̂_{k|k} and current input
        Derive U^(m) and update:
        ξ̂^(m)_{k|k} = U^(m)T x̂_{k|k}
        P_r^(m)(k|k) = U^(m)T P(k|k) U^(m),  Q_r^(m) = U^(m)T Q U^(m)
    end if
    Acquire new measurement y_k
    Update reduced state ξ̂^(m)_{k|k}
    Reconstruct full state estimate x̂_{k|k} = U^(m) ξ̂^(m)_{k|k}
    Compute e_L(k) and Δe_L(k)
end for
```

---

## 5.4 Field Implementation of the Proposed Framework

This section evaluates the performance of the proposed adaptive model reduction and state estimation framework through simulations applied to a large-scale agricultural field. The test site is a circular 26.4-hectare field located at Lethbridge Polytechnic's Research Farm, previously introduced in Chapter 4. Specifically, Field 3—depicted in Figure 4.5—serves as the study area for this chapter. The soil hydraulic parameters used here were obtained from the estimation methodology described in Chapter 4. While Chapter 4 focused on a single quadrant of Field 3, the present study extended the estimation procedure to the remaining quadrants.

The soil depth was set to 0.4 m and was evenly divided into 12 layers. The field radius of 290 m was discretized into 30 radial segments and 68 azimuthal segments, resulting in a total of 20,400 spatial nodes representing the system states. Pressure head measurements were collected at each sampling time using point sensors positioned at 90 locations across the field, covering both surface and subsurface depths. Weather data employed in this study were consistent with those used in Chapter 4. The initial pressure head values, used to simulate actual soil moisture at time $t_0$, were assigned distinct values for each quadrant of the field: −13.5, −14.0, −12.7, and −11.5 m, respectively.

The simulation was carried out over a period of 30 days using a sampling interval of $\Delta = 30$ minutes. The initial guess $\hat{x}_{0|0}$ for the soil moisture states was specified as −10.0, −12.0, −9.0, and −14.0 m for the four field quadrants, respectively. These initial guesses differed from the actual initial soil moisture values $x(0)$ assigned to each quadrant. The covariance matrices for the measurement noise and process disturbances were selected as $R = 0.08 \times I_{N_y=90}$ and $Q = 1.0 \times I_{N_x=20,400}$, where $I_{N_y=90}$ and $I_{N_x=20,400}$ denote identity matrices of dimensions $N_y = 90$ and $N_x = 20,400$, respectively. The initial state covariance matrix $P$ was defined such that all diagonal entries were set to 1.0, while all off-diagonal entries were assigned a value of $5 \times 10^{-5}$. The process disturbance and measurement noise signals were generated from zero-mean Gaussian distributions with variances of $10^{-7}$ and 0.8, respectively.

Three estimation schemes were considered in the simulation study to evaluate the effectiveness of the proposed approach. **Scheme I** implemented the proposed performance-triggered model reduction in conjunction with the adaptive EKF. **Scheme II** employed a reduced-order EKF based on a single, fixed reduced model that remained unchanged throughout the simulation period. **Scheme III** introduced a time-triggered adaptation mechanism, where the reduced model was periodically updated at fixed intervals, and the EKF was reconfigured accordingly. This scheme was intended to mitigate the impact of input disturbances without relying on performance-based triggers.

The error threshold and clustering threshold were set to $th_e = 40$ and $\delta_c = 1.0$, respectively. The tunable parameter $\Delta e_L$, used to track trends in model prediction error, was assigned a value of 0.05. For Schemes I and III, two different prediction horizon lengths were employed: $N_{f_d} = 250$ and $N_{f_d} = 350$, respectively.

### 5.4.1 Results and Discussion

Figure 5.2 illustrates the performance of the proposed performance-triggered adaptive EKF (Scheme I) for two different horizon lengths, $N_{fd} = 250$ and $N_{fd} = 350$. The top plots show the temporal variation in the percentage mean absolute error (%MAE), while the bottom plots indicate the time points at which model re-identification occurred, along with the associated model orders. At the beginning of the simulation, the reduced-order model operates with a low-dimensional representation, reflecting relatively homogeneous field conditions. As the season progresses and new inputs (e.g., irrigation, rainfall) are introduced, the system dynamics evolve, prompting increases in model order. For instance,

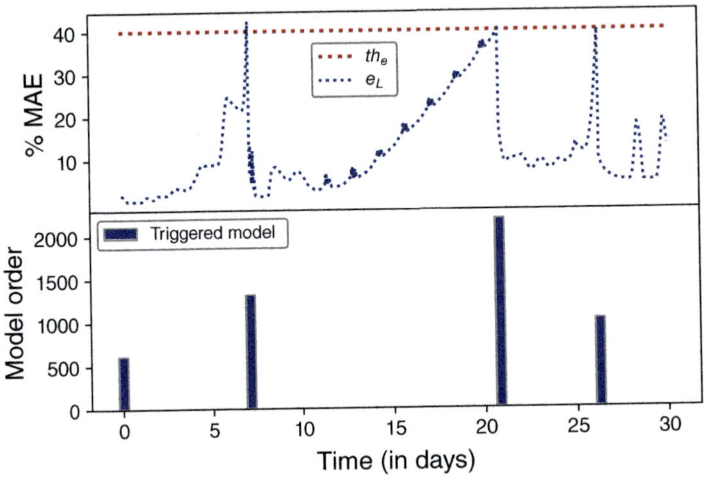


(a) $N_{fd} = 250$.

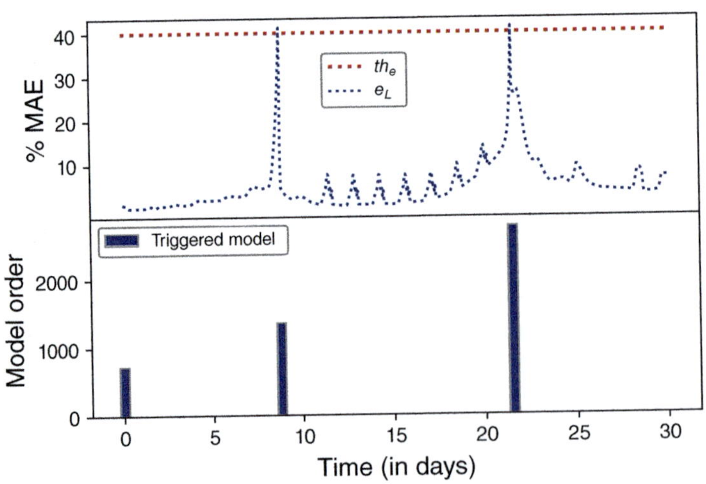


(b) $N_{fd} = 350$.

**Figure 5.2** Performance-triggered adaptive EKF (Scheme I): %MAE over time (top) and model re-identification instances with corresponding model orders (bottom). *Source*: [97]/John Wiley & Sons/CC BY 4.0.

the maximum model order observed is 2,778 for the longer horizon ($N_{f_d} = 350$), compared to 2,188 for the shorter horizon. Figure 5.2a shows an additional re-identification event due to the limited predictive window associated with the shorter horizon.

To provide a baseline comparison, Scheme II employs a nonadaptive model reduction approach. A fixed cluster generation threshold $\delta_c = 5.0$ is used to generate a single reduced model with order 1974 for the entire simulation period. While this configuration captures the dominant field dynamics, it lacks the flexibility to adapt to changes in system behavior over time.

To address this limitation, Scheme III introduces a time-triggered model re-identification strategy. Unlike Scheme II, the reduced model is updated periodically using fixed intervals. This approach helps mitigate the impact of uncertainties in weather forecasts and irrigation decisions. Figure 5.3 illustrates the performance of Scheme III for both $N_{f_d} = 250$

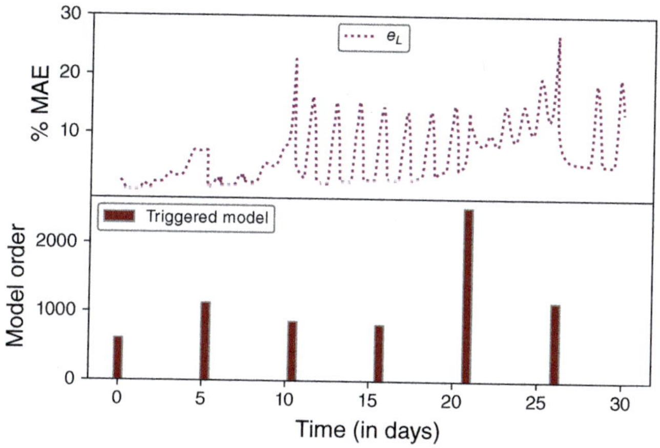


(a) $N_{fd} = 250$.

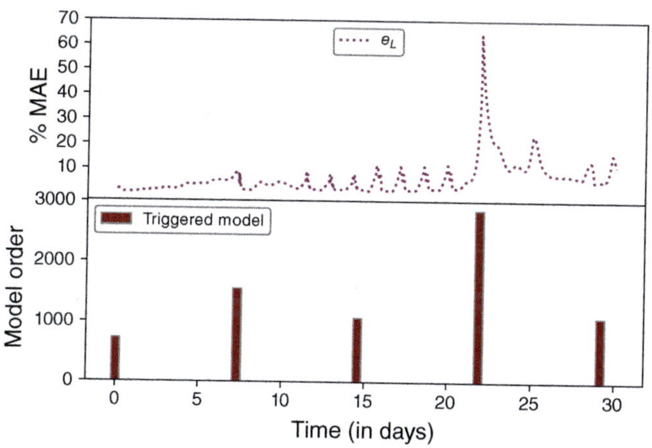


(b) $N_{fd} = 350$.

**Figure 5.3** Time-triggered adaptive EKF (Scheme III): %MAE over time (top) and model re-identification instances with corresponding model orders (bottom). *Source:* [97]/John Wiley & Sons/CC BY 4.0.

and $N_{fd} = 350$. Although the resulting estimation accuracy is comparable to that of Scheme I for a given horizon, Scheme III requires more frequent model updates, resulting in a higher number of model re-identifications. For example, the highest model orders reached are 2,528 and 2,868 for the short and long horizons, respectively. Overall, the results in Figure 5.4 reveal that the performance-triggered scheme (Scheme I) consistently achieves faster convergence and better control of estimation error compared to its nonadaptive counterpart. This is largely due to a lower estimation burden—each reduced model contains fewer states to estimate, while the number of available measurements remains unchanged. However, because Scheme I relies on multiple simplified models rather than a single full-order model, a slight increase in post-convergence error is observed due to accumulated model mismatch. Despite this minor trade-off, Scheme I offers a good balance between computational efficiency and estimation accuracy. Its flexibility in tuning (through parameters $N_{fd}$, $\delta_c$, and $th_e$) also makes it well-suited for practical implementation under variable field conditions.

### 5.4.2 Simulation Time Comparison Across Estimation Schemes

All simulations were conducted on a machine equipped with an Intel(R) Core(TM) i7-8700 CPU operating at 3.2 GHz and 24.0 GB of RAM.

As shown in Table 5.1, Scheme I, which combines adaptive clustering, model reduction, and recursive filtering, required approximately 5 seconds per sampling step. Scheme III incurred a slightly higher computational cost, about 5.5 seconds per step, primarily due to

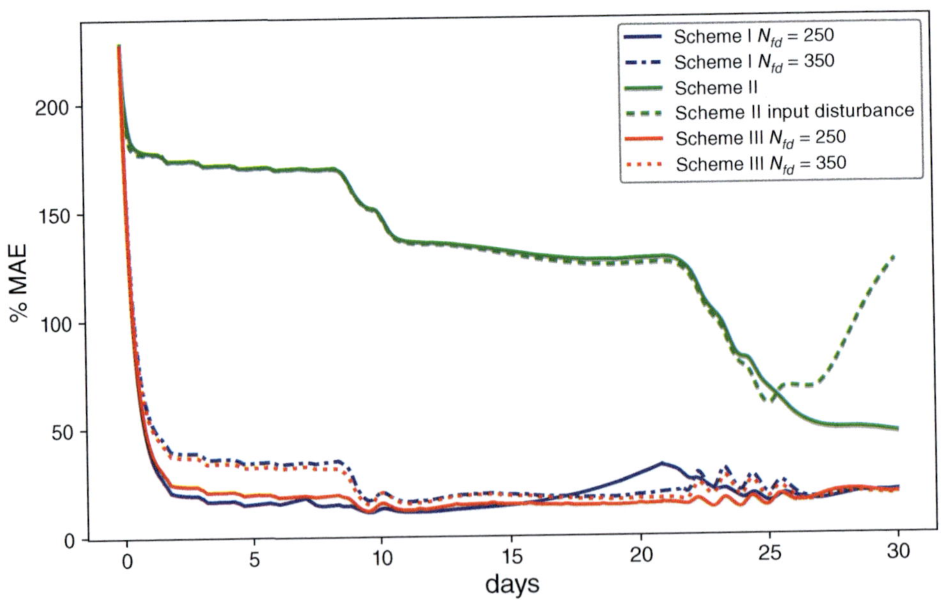


**Figure 5.4** Comparison of state estimation performance (%MAE) across the three schemes. *Source*: [97]/John Wiley & Sons/CC BY 4.0.

**Table 5.1** Average computation time per sampling step for different estimation schemes.

| Scheme | Average time |
|---|---|
| I with $N_{f_d}$ of 250 | 5.01 s |
| I with $N_{f_d}$ of 350 | 4.93 s |
| II | 2.20 s |
| III with $N_{f_d}$ of 250 | 5.52 s |
| III with $N_{f_d}$ of 350 | 5.48 s |

more frequent model updates. In contrast, Scheme II was computationally faster, with an average time of around 2.2 seconds per step. However, this comes at the cost of reduced estimation accuracy and greater sensitivity to input disturbances. Furthermore, when Scheme II employs a high-order reduced model to capture long-term dynamics, the computational burden increases substantially due to the need to evaluate a larger state transition matrix.

## 5.5 Conclusion

This chapter presented a performance-triggered adaptive model reduction framework integrated with an EKF for efficient soil moisture estimation in large-scale agricultural fields. The method dynamically adjusts the reduced-order model in response to estimation performance, enabling it to track evolving field dynamics while preserving computational efficiency. Simulation results based on a realistic field configuration showed that the proposed approach effectively balances estimation accuracy and computational cost. In comparison to a fixed, nonadaptive model, it demonstrated greater robustness to disturbances and better responsiveness to temporal variations in field conditions. While a time-triggered approach, in which model adaptation was performed at a fixed time interval, exhibited comparable accuracy, it required more frequent model updates, which resulted in higher computation times. In summary, the proposed framework significantly reduces the dimensionality of the estimation problem without compromising accuracy, making it a scalable and practical solution for precision irrigation applications.

# 6

# Mixed-integer Model Predictive Control for Irrigation Scheduling

## 6.1 Introduction

This chapter builds on prior model predictive control (MPC) approaches for irrigation by introducing a mixed-integer MPC framework tailored to daily scheduling across diverse agricultural fields. In contrast to conventional MPC-based irrigation strategies, which often omit discrete decision-making, the proposed method explicitly models both the timing and quantity of irrigation. Binary variables specify which days irrigation should occur, while continuous variables define the water volumes to be applied. The scheduler integrates a process-based model of the soil–crop–atmosphere system and is designed to achieve three key objectives: keeping soil moisture within an optimal range, minimizing total water use, and reducing costs associated with the irrigation event.

The initial implementation of the mixed-integer MPC scheduler focuses on homogeneous fields, where soil and crop characteristics are spatially uniform. To ease the computational load associated with solving the resulting optimization problem, the soil moisture dynamics are approximated using long short-term memory (LSTM) models. This surrogate is trained on datasets obtained from open-loop simulations of the Richards' equation, which serves as the foundational model for the soil–crop–atmosphere interactions. The scheduler's performance is then tested in this uniform field context to evaluate how well it achieves the specified irrigation goals.

The mixed-integer MPC framework is later adapted to address spatial heterogeneity by introducing management zones, each reflecting localized differences in soil characteristics and crop types. To retain computational efficiency, separate LSTM models are trained for each zone, enabling accurate yet tractable modeling of soil moisture dynamics. Additionally, the formulation leverages a smooth approximation of binary variables using the sigmoid function—a method previously applied in domains such as transmission expansion planning [98] and power loss reduction [99]. The chapter ends with a case study involving a spatially variable field, demonstrating the scheduler's ability to maintain target moisture levels across zones while conserving water and minimizing operational costs. The insights presented build on earlier work in [100].

*Precision Irrigation for Agriculture: Integrating Machine Learning and Optimal Control Strategies*,
First Edition. Bernard Twum Agyeman and Jinfeng Liu.

## 6.2 Daily Irrigation Scheduling Under Uniform Field Conditions

A field is considered homogeneous when its primary characteristics such as soil composition, crop type, and elevation remain uniform across the entire area. An example of such a field is illustrated in Figure 6.1. In this context, the daily irrigation scheduling problem reduces to determining, for each day within a specified planning horizon, whether irrigation should be applied and, if so, the appropriate amount. The spatial uniformity of the field allows for a single irrigation amount to satisfy crop water needs across the entire area.

To capture irrigation timing decisions, a binary decision variable is defined for each day in the scheduling horizon. This variable takes the value 0 or 1. When it assumes a value of 0, no water is applied on that day. When it is 1, irrigation is scheduled, and the associated irrigation amount must be computed. Accordingly, the task of scheduling irrigation on a daily basis in a uniform field can be formally defined as:

**Given:**

1. **Scheduling Horizon**: The total number of days over which irrigation decisions are to be made.
2. **Weather Forecasts**: Daily predictions of reference evapotranspiration and precipitation. The former represents potential atmospheric water loss through evaporation and transpiration, while the latter accounts for natural water input to the field.
3. **Crop Information**: Captured by the crop coefficient, which scales reference evapotranspiration to reflect crop-specific water requirements at different phenological stages.

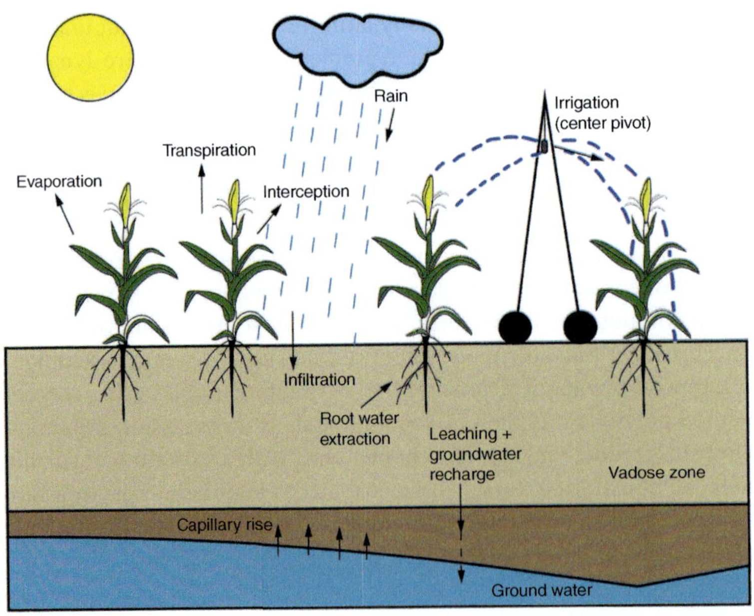


**Figure 6.1** An agro-hydrological system characterized by uniform soil properties and a single crop type. *Source*: [42]/MDPI/CC BY 4.0.

4. **Soil Moisture Content**: The moisture profile within the crop's rooting zone at the start of the horizon, used to estimate the additional water needed to maintain optimal soil conditions.

**Determine:**

1. **Irrigation Timing**: Days on which irrigation is to be carried out.
2. **Irrigation Amounts**: The quantity of water to apply each day, set to zero on non-irrigation days and to a positive value on scheduled irrigation days, in accordance with the irrigation timing.

### 6.2.1 Soil Moisture Model Development

Infiltration events, such as rainfall and irrigation, are the primary drivers of soil moisture dynamics in the soil–crop–atmosphere system. In fields with spatial homogeneity in crop and soil attributes, the 1D Richards' equation in the vertical direction offers a sufficient framework for modeling moisture movement through the soil profile. For completeness, the 1D form of the Richards' equation, originally introduced in Chapter 2, is restated below:

$$C(\psi)\frac{\partial \psi}{\partial t} = \frac{\partial}{\partial z}\left[K(\psi)\left(\frac{\partial \psi}{\partial z}+1\right)\right] - \rho(\psi)\mathcal{R}\,(K_c, \mathrm{ET}_0, z_r) \tag{6.1}$$

The numerical solution of Equation (6.1) is performed subject to the following conditions:

$$\psi(t=0) = \psi_{\text{init}} \qquad \text{(initial condition)} \tag{6.2}$$

$$\left.\frac{\partial(\psi+z)}{\partial z}\right|_{z=\mathrm{H}_z} = 1 \qquad \text{(bottom boundary condition)} \tag{6.3}$$

$$\left.\frac{\partial \psi}{\partial z}\right|_{z=0} = -1 - \frac{u^{\text{irr}} - \mathrm{EV}}{K(\psi)} \qquad \text{(top boundary condition)} \tag{6.4}$$

where $\psi_{\text{init}}$ represents the initial distribution of pressure head along the soil column of depth $\mathrm{H}_z$. Its dimensionality corresponds to the number of spatial discretization points defined within $\mathrm{H}_z$. In Equation (6.4), the term $u^{\text{irr}}$ denotes the irrigation amount, while EV represents the evaporation.

Using the numerical method described in Section 2.6.1 of Chapter 2, the 1D Richards' equation is transformed into a state-space representation:

$$x_{k+1} = \mathcal{F}(x_k, u_k, \boldsymbol{\phi}) + \omega_k \tag{6.5}$$

$$y_k = \mathcal{H}(x_k, \boldsymbol{\phi}) + v_k \tag{6.6}$$

Within the state-space framework above, the state vector $x_k \in \mathbb{R}^{N_x}$ captures the distribution of capillary pressure head across $N_x$ discrete points along the soil column $\mathrm{H}_z$. The parameter vector $\boldsymbol{\phi}$ aggregates the five primary soil hydraulic parameters necessary for solving Equation (6.1), and is defined as $\boldsymbol{\phi} := [K_s, \theta_s, \theta_r, \alpha, n]$. The functions $\mathcal{F}$ and $\mathcal{H}$, appearing in Equations (6.5) and (6.6), represent the process model and the measurement model, respectively. The term $u_k$ aggregates the inputs of the state-space model, and it includes irrigation, rainfall, reference evapotranspiration, and the crop coefficient. The term $\omega_k$ represents the process uncertainty while $v_k$ denotes the measurement noise. The output vector $y_k \in \mathbb{R}^{N_y}$ reflects the volumetric water content at all $N_x$ nodes, implying $N_y = N_x$. Hence,

Equation (6.6) extends Equation (2.9) by expressing a generalized relationship between pressure head and soil moisture across the entire spatial domain.

### 6.2.2 Surrogate Model Development

Although Equation (6.1) accurately describes soil moisture dynamics in homogeneous fields, its nonlinear nature and reliance on complex soil hydraulic functions make it computationally demanding for use in MPC-based irrigation scheduling. To address this issue, surrogate modeling techniques can be used to approximate the system's behavior. These simplified representations retain essential dynamic features while significantly reducing the computational burden, making them more suitable for real-time control applications.

Surrogate models are designed to mimic the response of complex systems, and in agro-hydrological applications where temporal dynamics are critical, LSTM networks serve as a suitable modeling choice. LSTM networks are capable of learning temporal dependencies from sequential data, making them well-suited for predicting the evolution of soil moisture in response to inputs such as irrigation, rainfall, and evapotranspiration. A detailed explanation of LSTM networks is provided in Section 3.1.1.2 of Chapter 3.

Developing a surrogate model for the field begins with the generation of an appropriate training dataset. A common approach is to simulate the system's behavior using open-loop runs of the Richards' equation. These runs yield time-series data that reflect soil moisture responses under different environmental and irrigation scenarios, enabling the LSTM network to learn and generalize the dynamics of the underlying system.

This section presents a procedure for constructing an LSTM model suited to homogeneous agricultural fields. It first details how training data are derived from simulations of the Richards' equation, followed by a description of a network architecture designed to effectively approximate soil moisture dynamics with adequate precision.

#### 6.2.2.1 Dataset Generation

To develop the LSTM model under homogeneous field conditions, a training dataset is generated by numerically simulating Equations (6.5) and (6.6) using randomized inputs $u_k$ and diverse initial states $x_0$. To improve the generalization capability of the surrogate model and capture inherent system uncertainties, controlled noise is incorporated during the data generation phase. The variation in both inputs and initial conditions allows the dataset to capture a wide range of soil moisture responses under differing environmental and management scenarios.

In the context of calculating irrigation schedules, the key output variable is the root zone soil moisture content, $\theta^{\mathrm{RZ}}$. This quantity is typically treated as a scalar derived from the full soil moisture profile $y$, which varies with depth. Although weighted averaging methods are commonly used to estimate root zone moisture in practice, a simplified approach is employed in this chapter. In particular, $\theta^{\mathrm{RZ}}$ is approximated using the moisture value at a single representative depth. This is defined as:

$$\theta^{\mathrm{RZ}} := \mathcal{M}y \tag{6.7}$$

where $\mathcal{M} \in \mathbb{R}^{1\times N_y}$ is a row vector containing zeros at all positions except the one corresponding to the chosen root zone depth.

Following the completion of the simulations, Equation (6.7) is applied at each time step to compute the root zone soil moisture values, which are then compiled to form the training dataset.

#### 6.2.2.2 LSTM Architecture

The surrogate model aims to provide daily predictions of the root zone volumetric moisture content, providing the value for the following day, $\theta^{\text{RZ}}_{t+1}$. This prediction is based on a sequence of current and past input data points, spanning from time step $t$ back to $t-l$, where $l$ represents the sequence length. The inputs supplied to the LSTM network consist of $\theta^{\text{RZ}}$, $u^{\text{irrig}}$ (the combined effect of rainfall and applied irrigation), crop coefficient $K_c$, and reference evapotranspiration $\text{ET}_0$. The mathematical representation of the surrogate model is given by:

$$\hat{\theta}^{\text{RZ}}_{t+1} = \mathcal{F}_{\text{LSTM}}(\{\gamma\}^{t}_{t-l}, \eta) \tag{6.8}$$

where $\{\gamma\}^{t}_{t-l} := [\gamma_{t-l}, .., \gamma_t]$, $\gamma \in [\hat{\theta}^{\text{RZ}}, u^{\text{irrig}}, K_c, \text{ET}_0]$, and $\eta$ represents the set of trainable parameters learned by the network.

### 6.2.3 Scheduler Formulation

The daily irrigation scheduling problem in a homogeneous agricultural setting can be viewed as a modified form of the classical warehouse location problem. The irrigation system assumes the role of the warehouse, and the decision to initiate irrigation on a given day parallels the decision to operate a warehouse facility. Fixed irrigation costs correspond to the fixed costs of opening the warehouse, while the volume of water applied reflects the variable operational cost associated with the warehouse's daily operations.

The core objective of the scheduling model is to regulate root zone soil moisture within some predetermined optimal bounds. This is achieved by incorporating penalty terms into the objective function that discourage significant violations of the bounds. The optimization is subject to a process model that governs the evolution of soil moisture dynamics in the homogeneous field. Additionally, soft constraints are imposed to ensure that the predicted soil moisture remains within relaxed limits of the optimal bounds over the scheduling horizon.

Alongside moisture regulation, the scheduler also seeks to minimize total fixed and variable irrigation costs. This particular objective mirrors that of the warehouse location problem, where total costs are minimized while meeting service demand. In the irrigation context, this translates to delivering sufficient water to satisfy crop requirements without exceeding the limits of the irrigation implementing equipment.

The combination of these objectives, along with their associated constraints, results in a mixed-integer MPC formulation with zone control. To reduce computational complexity, the framework integrates an LSTM-based surrogate model, as defined in Equation (6.8), to approximate the 1D Richards' equation. Given day $d$ as the starting point of a scheduling horizon of length $N$, and an initial soil moisture vector $y$ in the field, the MPC problem is denoted by $\mathbb{P}_{\mathrm{N}}(y, d)$ and is formulated as:

$$\min_{\bar{\epsilon}, \underline{\epsilon}, \boldsymbol{u}^{\text{irrig}}, \boldsymbol{c}} \sum_{k=d+1}^{d+N} \left[\bar{Q}\bar{\epsilon}_k^2 + \underline{Q}\underline{\epsilon}_k^2\right] + \sum_{k=d}^{d+N-1} R_c c_k + \sum_{k=d}^{d+N-1} R_u u_k^{\text{irrig}} \tag{6.9a}$$

s.t.

$$\theta_{k+1}^{\text{RZ}} = \mathcal{F}_{\text{LSTM}}(\{\gamma\}_{k-l}^{k}, \eta) \qquad \forall k \in [d, d+N-1] \tag{6.9b}$$

$$\theta_d^{\text{RZ}} = \mathcal{M}y(d) \tag{6.9c}$$

$$\underline{\nu} - \underline{\epsilon}_k \leq \theta_k^{\text{RZ}} \leq \bar{\nu} + \bar{\epsilon}_k, \qquad \forall k \in [d+1, d+N] \tag{6.9d}$$

$$c_k \underline{u}^{\text{irrig}} \leq u_k^{\text{irrig}} \leq c_k \bar{u}^{\text{irrig}}, \qquad \forall k \in [d, d+N-1] \tag{6.9e}$$

$$c_k = \{0, 1\}, \qquad \forall k \in [d, d+N-1] \tag{6.9f}$$

$$\underline{\epsilon}_k \geq 0, \quad \bar{\epsilon}_k \geq 0, \qquad \forall k \in [d+1, d+N] \tag{6.9g}$$

where $k \in \mathbb{Z}^+$. The vectors $\bar{\epsilon} := [\bar{\epsilon}_{d+1}, ..., \bar{\epsilon}_{d+N}]$ and $\underline{\epsilon} := [\underline{\epsilon}_{d+1}, ..., \underline{\epsilon}_{d+N}]$ contain the upper and lower slack variables, respectively, introduced to relax the soil moisture target zone with bounds of $(\underline{\nu}, \bar{\nu})$. The vector $\boldsymbol{c} := [c_d, ..., c_{d+N-1}]$ represents the binary decisions for irrigation timing and $\boldsymbol{u}^{\text{irrig}} := [u_d^{\text{irrig}}, ..., u_{d+N-1}^{\text{irrig}}]$ contains the corresponding irrigation amounts. Equation (6.9b) defines the LSTM model and the term $\gamma \in [\theta^{\text{RZ}}, u^{\text{irrig}}, K_c, \text{ET}_0]$ represents an aggregation of its inputs. The notation $\{\gamma\}_{k-l}^{k} := [\gamma_{k-l}, ..., \gamma_k]$ refers to the input sequence of length $l$. The variables $K_c$ and $\text{ET}_0$ are treated as uncontrolled inputs, and are obtained from forecasted weather conditions.

The parameters $\underline{Q}$ and $\bar{Q}$ represent the per-unit penalties applied when the root zone soil moisture content falls below or exceeds the defined target range, respectively. The terms $R_u$ and $R_c$ correspond to the unit cost of water and the cost incurred from initiating the irrigation event, respectively. The objective function integrates three goals. First, it aims to maintain the root zone soil moisture within the desired range by minimizing deviations from the target zone using the penalty expression $\sum_{k=d+1}^{d+N} \left[\bar{Q}\bar{\epsilon}_k^2 + \underline{Q}\underline{\epsilon}_k^2\right]$. Second, it seeks to reduce the fixed irrigation cost through the expression $\sum_{k=d}^{d+N-1} R_c c_k$. Third, it minimizes total water usage through the expression $\sum_{k=d}^{d+N-1} R_u u_k^{\text{irrig}}$.

The initial state of the system, defined by Equation (6.9c), is established using the spatial soil moisture profile observed at the beginning of $N$. Constraint Equation (6.9e) governs the permissible irrigation amount for each day. When no irrigation is planned on day $k$ (i.e., $c_k = 0$), the irrigation input must be zero. Conversely, if irrigation is scheduled (i.e., $c_k = 1$), the volume of applied water must fall within the allowable range, bounded by $\underline{u}^{\text{irrig}}$ and $\bar{u}^{\text{irrig}}$.

### 6.2.4 Case Study

The performance of $\mathbb{P}_{\mathrm{N}}(y, d)$ is examined in a uniform field setting represented by a 0.60-m-deep loamy soil profile cultivated with wheat at its initial developmental stage. This simplified configuration, defined by uniform soil and crop characteristics, provides an appropriate context for evaluating the scheduler's ability to maintain root zone moisture within prescribed limits and promote efficient irrigation.

The evaluation process begins with the construction of an LSTM model to approximate soil moisture behavior under these conditions. This is followed by a description of the simulation setup and a detailed assessment of the results obtained from applying the scheduling framework.

#### 6.2.4.1 Model Training

Prior to deploying the scheduler, an LSTM network was trained to forecast the root zone soil moisture content in the field. The data employed for the model development were generated using the 1D Richards' equation, parameterized for loamy soil (refer to Table 6.1). The field was divided into 30 equal layers in the vertical direction, and the model was simulated at a temporal resolution of 6 minutes. The initial condition was set by assigning a uniform pressure head of −0.180 m throughout the soil column.

To generate a dataset reflecting a wide range of soil moisture dynamics, the Richards' equation was numerically solved under varying input conditions. Input values were randomly sampled from the following intervals: irrigation amount between 0.33 and 3.56 cm/day, rainfall between 1.04 and 7.0 mm, reference evapotranspiration between 1.04 and 3.0 mm, and crop coefficient values ranging from 0.50 to 0.88. Irrigation was applied at regular intervals, with the time between consecutive events randomly drawn from a uniform distribution spanning 2–6 days. To reflect process variability, zero-mean Gaussian noise with a standard deviation of 0.00005 m was added to the simulation data.

Soil moisture at the 0.5-m depth was selected as a representative measure of root zone moisture. After computing the root zone moisture values, the resulting dataset was resampled to a daily resolution, producing a dataset of 56,000 entries. The architecture consisted of two hidden layers, each comprising 200 LSTM units. A time lag of 5 days was employed for the model development.

#### 6.2.4.2 Simulation Settings

The bounds for the soil moisture range were defined as 0.255 $m^3/m^3$ (lower limit) and 0.273 $m^3/m^3$ (upper limit). Deviations from this range were penalized using cost parameters $\bar{Q} = \underline{Q} = 9,000$. The cost of initiating the irrigation event was set to $R_c = 50$, and the cost per unit of water was specified as $R_u = 20$. The initial root zone moisture was assumed to be known and was set to 0.266 $m^3/m^3$.

The actual system dynamics were represented using the 1D Richards' equation, configured to model a 0.6-m-deep loamy soil profile. At each time step, the scheduler computed

**Table 6.1** Soil hydraulic properties for loamy and loamy sand.

| Soil type | $K_s$ (m/s) | $\theta_s$ ($m^3/m^3$) | $\theta_r$ ($m^3/m^3$) | $\alpha$ (1/m) | $n$ (−) |
|---|---|---|---|---|---|
| Loam | $2.889 \times 10^{-6}$ | 0.430 | 0.078 | 3.600 | 1.56 |
| Loamy sand | $4.053 \times 10^{-5}$ | 0.410 | 0.057 | 12.40 | 2.28 |

*Source:* Adapted from [86].

an optimal irrigation sequence over a 14-day horizon, but only the irrigation amount corresponding to the first day's irrigation amount was implemented. This procedure was repeated daily, with updated soil moisture observations taken from a depth of 0.50 m serving as the new initial condition. The simulation spanned a total of 20 days, during which the crop coefficient was maintained at a constant value of 0.50. The underlying optimization problem was formulated as a mixed-integer program and solved using the BONMIN solver [77]. Figure A.2 in Appendix A.2 shows the reference evapotranspiration data used in this experiment.

#### 6.2.4.3 Simulation Results

Figure 6.2a displays a comparison between the LSTM model's one-day-ahead predictions and the actual root zone soil moisture content from the test set. The plots demonstrate that the trained LSTM model closely tracks the observed moisture levels and captures the overall behavior of soil moisture variation. The model achieved a root-mean-square error (RMSE) of 0.0015 $m^3/m^3$ and an $R^2$ value of 0.99, indicating good predictive accuracy.

Figure 6.2b shows multistep-ahead forecasts resulting from a recursive application of the identified LSTM network. These values were obtained by using the model's own outputs as inputs for successive time steps. The model continues to perform well over multiple steps, with an RMSE of 0.0022 $m^3/m^3$ and an $R^2$ of 0.97. This predictive capability is particularly

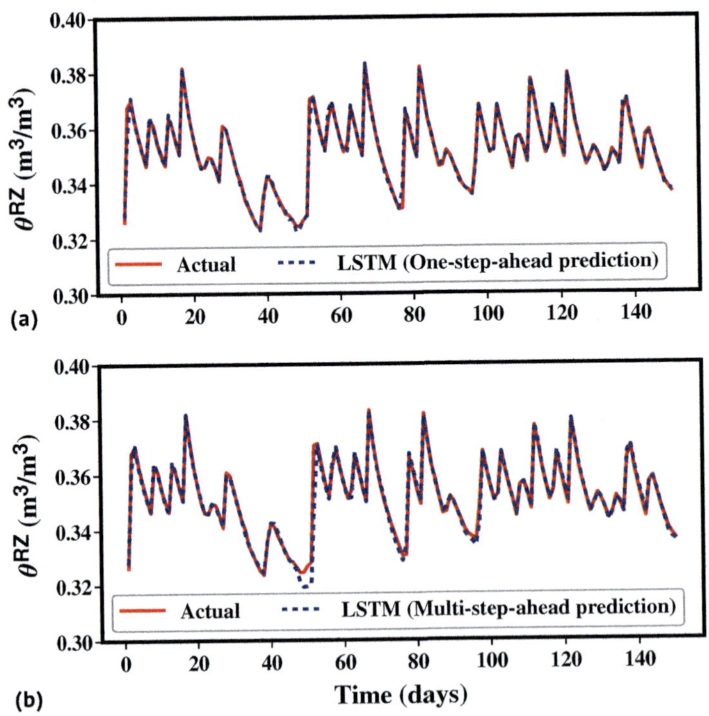


**Figure 6.2** Comparison of actual (red solid line) and predicted (blue dash-dot line) root zone soil moisture content using the test dataset: (a) one-day-ahead prediction and (b) multistep-ahead prediction.

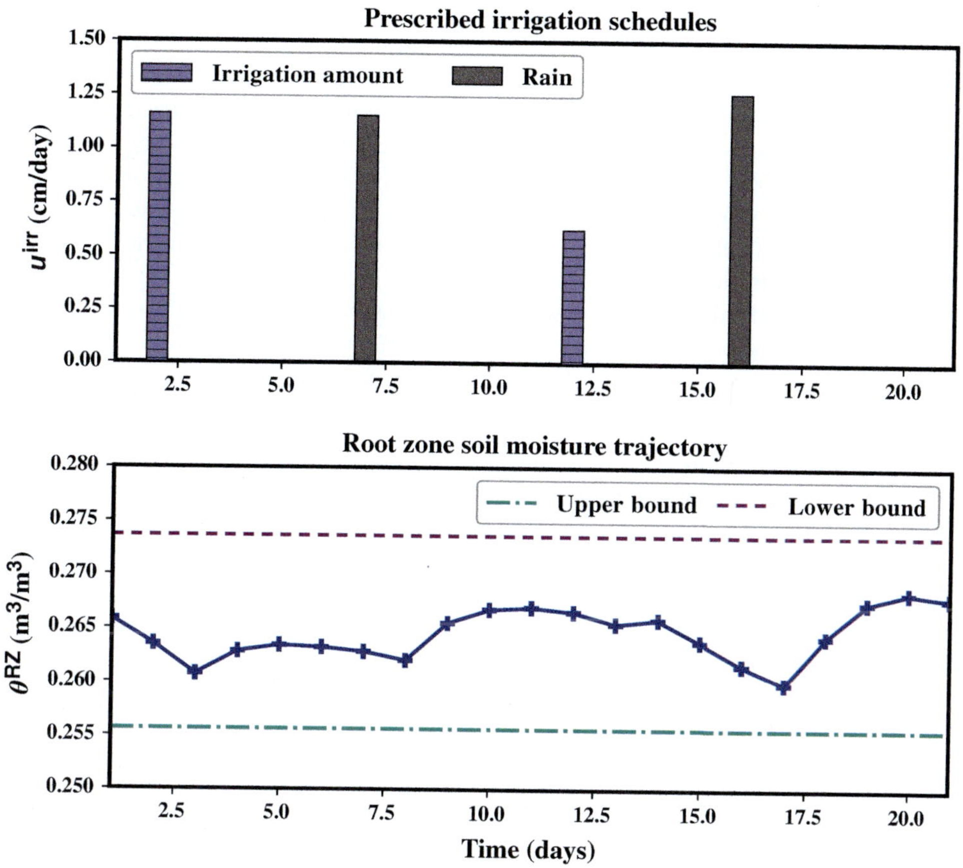


**Figure 6.3** Prescribed irrigation and resulting root zone soil moisture trajectories for the homogeneous field.

advantageous for the MPC framework, which relies on multistep forecasts of soil moisture to generate optimal irrigation schedules.

Figure 6.3 presents the irrigation schedules alongside the resulting root zone soil moisture profiles and recorded precipitation on days 7 and 16. Irrigation was triggered on days 2 and 12, with application depths of 1.16 and 0.6 cm/day, respectively, to ensure that soil moisture remained within the specified target range.

## 6.3 Irrigation Scheduling in Spatially Heterogeneous Fields

Agricultural fields typically show significant spatial heterogeneity due to interacting chemical, physical, and biological occurrences within the soil profile. This variability poses challenges for uniform irrigation practices, as a one-size-fits-all approach often results in inefficient water distribution. As a result, the mixed-integer MPC strategy described in

Section 6.2, which assumes field homogeneity, is insufficient for effective water management in spatially variable conditions. Achieving optimal water use under such circumstances requires a scheduling framework that reflects spatial variability.

A widely adopted solution is to partition the field into multiple management zones (MZs)—subsections of the field characterized by relatively uniform soil and crop attributes. The delineation of MZs typically considers factors such as soil texture, elevation, drainage characteristics, and empirical field knowledge. Figure 6.4 provides a schematic of a spatially heterogeneous field comprising two different soil types and two crop varieties. Based on these distinctions, the field is divided into two MZs.

In these settings, the daily irrigation scheduling process must determine the specific days on which irrigation should be applied and the corresponding irrigation amounts. This leads to the following formulation of the scheduling problem for spatially heterogeneous fields:

**Given:**

1. **Scheduling Horizon**: Days over which schedules ought to be calculated.
2. **Management Zones**: The field is divided into $M$ distinct zones.
3. **Weather Predictions**: Predictions of daily values of rainfall and reference evapotranspiration across the horizon.
4. **Crop Information**: Crop coefficient values of the crops in each zone.
5. **Soil Moisture Content**: Soil moisture profile within the rooting depth across each MZ at beginning of the horizon.

**Determine:**

1. **Irrigation Timing**: The days within the scheduling horizon on which irrigation should occur. Timing must be consistent across all zones—i.e., if irrigation is applied on a given day, it must be applied in all zones.
2. **Irrigation Amounts**: The amount of water to be applied in each zone on scheduled irrigation days. On non-irrigation days, these amounts are zero across all zones; on irrigation days, each zone receives a specified amount tailored to its needs.

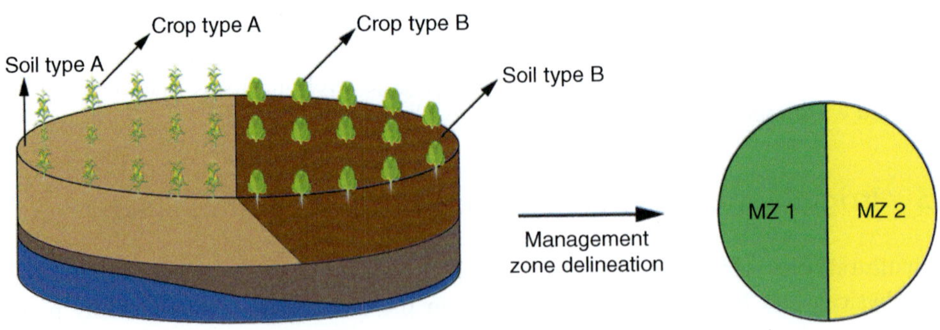


**Figure 6.4** Illustration of a field exhibiting spatial variability in soil and crop characteristics. The field is partitioned into two distinct MZs, each characterized by uniform properties.

### 6.3.1 Soil Moisture Modeling

The influence of one MZ on another is typically minimal, in large-scale fields. This limited interaction permits each MZ to be treated as an independent unit. Consequently, the modeling approach used for homogeneous fields can be applied individually to each zone.

In the case of spatially heterogeneous fields composed of multiple MZs, the state-space formulation of the 1D Richards' equation, originally defined for a single homogeneous domain in Equations (6.5) and (6.6), can be extended. The collective dynamics are represented using a stacked state vector $\boldsymbol{X}_k \in \mathbb{R}^{N_x \times M}$ and an output vector $\boldsymbol{Y}_k \in \mathbb{R}^{N_y \times M}$, where each column captures the respective states and outputs for one of the $M$ MZs at time step $k$. This extended state-space representation is defined as:

$$\boldsymbol{X}_{k+1} = \begin{bmatrix} x_{k+1,1} \\ x_{k+1,2} \\ \vdots \\ x_{k+1,M} \end{bmatrix} = \begin{bmatrix} \mathcal{F}\left(x_{k,1}, u_{k,1}, \boldsymbol{\phi}_1\right) + \omega_{k,1} \\ \mathcal{F}\left(x_{k,2}, u_{k,2}, \boldsymbol{\phi}_2\right) + \omega_{k,2} \\ \vdots \\ \mathcal{F}\left(x_{k,M}, u_{k,M}, \boldsymbol{\phi}_M\right) + \omega_{k,M} \end{bmatrix} \tag{6.10}$$

$$\boldsymbol{Y}_k = \begin{bmatrix} y_{k,1} \\ y_{k,2} \\ \vdots \\ y_{k,M} \end{bmatrix} = \begin{bmatrix} \mathcal{H}(x_{k,1}, \boldsymbol{\phi}_1) + v_{k,1} \\ \mathcal{H}(x_{k,2}, \boldsymbol{\phi}_2) + v_{k,2} \\ \vdots \\ \mathcal{H}(x_{k,M}, \boldsymbol{\phi}_M) + v_{k,M} \end{bmatrix} \tag{6.11}$$

where the term $\boldsymbol{\phi}_i$ is the hydraulic properties specific to the soil within the $i$th zone. The vectors $x_{k,i}$, $u_{k,i}$, and $y_{k,i}$ correspond to the state, input, and output of zone $i$ at time step $k$, respectively. The terms $\omega_{k,i}$ and $v_{k,i}$ capture the process and measurement uncertainties, respectively.

To enable efficient MPC-based irrigation scheduling, the moisture dynamics in each MZ is approximated with a surrogate model. Particularly, an LSTM network is trained for each MZ, in accordance with the methodology in Section 6.2.2. Each network independently generates one-day-ahead forecasts of root zone soil moisture for the MZ it represents:

$$\hat{\theta}^{\text{RZ}}_{t+1,1} = \mathcal{F}_{\text{LSTM},1}(\{\gamma_1\}^t_{t-l}, \eta_1) \tag{6.12}$$

$$\hat{\theta}^{\text{RZ}}_{t+1,2} = \mathcal{F}_{\text{LSTM},2}(\{\gamma_2\}^t_{t-l}, \eta_2) \tag{6.13}$$

$$\vdots$$

$$\hat{\theta}^{\text{RZ}}_{t+1,M} = \mathcal{F}_{\text{LSTM},M}(\{\gamma_M\}^t_{t-l}, \eta_M) \tag{6.14}$$

### 6.3.2 Scheduler Formulation

The mixed-integer MPC framework is modified to incorporate all $M$ MZs of the field. By doing so, the scheduler is equipped to generate zone-specific schedules that reflect the unique soil moisture profiles and crop water needs of each zone, ultimately enhancing irrigation efficiency. Given $d$, $N$, and $\boldsymbol{Y}$, the extended scheduling problem is represented by $\mathbb{P}_{\mathrm{M,N}}(\boldsymbol{Y}, d)$ and is formulated as:

$$\min_{\bar{\boldsymbol{E}}, \underline{\boldsymbol{E}}, \boldsymbol{U}^{\mathrm{irrig}}, \boldsymbol{c}} \sum_{j=1}^{M} \sum_{k=d+1}^{d+N} \left[\bar{Q}(j)\,\bar{\epsilon}_{jk}^2 + \underline{Q}(j)\,\underline{\epsilon}_{jk}^2\right] + R_c \sum_{k=d}^{d+N-1} c_k + R_u \sum_{j=1}^{M} \sum_{k=d}^{d+N-1} u_{jk}^{\mathrm{irrig}}$$

s.t.

$$\theta_{j(k+1)}^{\mathrm{RZ}} = \mathcal{F}_{\mathrm{LSTM}}^{j}(\{\gamma_j\}_{k-l}^{k}, \eta_j) \qquad \forall j \in \tilde{M}, \qquad \forall k \in [d, d+N-1] \tag{6.15a}$$

$$\theta_{yd}^{\mathrm{RZ}} = \mathcal{M} y_j(d) \qquad \forall j \in \tilde{M} \tag{6.15b}$$

$$\underline{\nu}_j - \underline{\epsilon}_{jk} \leq \theta_{jk}^{\mathrm{RZ}} \leq \bar{\nu}_j + \bar{\epsilon}_{jk}, \qquad \forall j \in \tilde{M}, \qquad \forall k \in [d+1, d+N] \tag{6.15c}$$

$$c_k \underline{u}_j^{\mathrm{irrig}} \leq u_{jk} \leq c_k \bar{u}_j^{\mathrm{irrig}}, \qquad \forall j \in \tilde{M}, \qquad \forall k \in [d, d+N-1] \tag{6.15d}$$

$$c_k \in \{0, 1\}, \qquad \forall k \in [d, d+N-1] \tag{6.15e}$$

$$\underline{\epsilon}_{jk} \geq 0, \quad \bar{\epsilon}_{jk} \geq 0, \qquad \forall j \in \tilde{M}, \qquad \forall k \in [d+1, d+N] \tag{6.15f}$$

where $j, k \in \mathbb{Z}^+$, and $\tilde{M} := [1, 2, ..., M]$ represents the index set of all MZs. The vectors $\bar{\boldsymbol{E}} := [\bar{\epsilon}_1, ..., \bar{\epsilon}_M]$ and $\underline{\boldsymbol{E}} := [\underline{\epsilon}_1, ..., \underline{\epsilon}_M]$ capture the upper and lower slack variables for all MZs, respectively. Similarly, $\boldsymbol{U}^{\mathrm{irrig}} := [\boldsymbol{u}_1^{\mathrm{irrig}}, ..., \boldsymbol{u}_M^{\mathrm{irrig}}]$ aggregates the irrigation amounts across all $M$ MZs.

In essence, $\mathbb{P}_{\mathrm{M,N}}(\boldsymbol{Y}, d)$ applies $\mathbb{P}_{\mathrm{N}}(y, d)$ to all $M$ MZs. While each zone follows its own version of $\mathbb{P}_{\mathrm{N}}(y, d)$, a global constraint (Equation 6.15d) is imposed to enforce a consistent timing across the zones of the field. That is, although irrigation quantities may vary across MZs, the decision to irrigate on any given day must be applied consistently throughout the entire field.

#### 6.3.2.1 Approximating the Binary Values with a Sigmoid Function

Although the BONMIN solver was suitable for addressing the mixed-integer scheduling problem in homogeneous fields, applying the same approach to spatially variable fields with several MZs introduces considerable computational complexity. To enhance the tractability of $\mathbb{P}_{\mathrm{M,N}}(\boldsymbol{Y}, d)$, the binary decision variables are approximated using a continuous relaxation through the sigmoid function $\omega(l)$, which is expressed as:

$$\omega(l) = \frac{1}{1 + e^{-\beta l}} \tag{6.16}$$

where $\beta$ controls the steepness of $\omega(\cdot)$, and $l$ is a continuous-valued input. Incorporating Equation (6.16) into the scheduling framework transforms the problem into a nonlinear programming (NLP) formulation, denoted by $\mathbb{P}_{\text{M,N}}^{\text{SIG}}(\boldsymbol{Y}, d)$:

$$\min_{\bar{E}, \underline{E}, U^{\text{irrig}}, l} \sum_{j=1}^{M} \sum_{k=d+1}^{d+N} [\bar{Q}(j)\bar{\epsilon}_{jk}^2 + \underline{Q}(j)\underline{\epsilon}_{jk}^2] + R_c \sum_{k=d}^{d+N-1} \omega(l_k) + R_u \sum_{j=1}^{M} \sum_{k=d}^{d+N-1} u_{jk}^{\text{irrig}}$$

s.t.

$$\theta_{j(k+1)}^{\text{RZ}} = \mathcal{F}_{\text{LSTM}}^{j}(\{\gamma_j\}_{k-l}^{k}, \eta_j) \quad \forall j \in \tilde{M}, \quad \forall k \in [d, d+N-1] \tag{6.17a}$$

$$\theta_{yd}^{\text{RZ}} = \mathcal{M}y_j(d) \quad \forall j \in \tilde{M} \tag{6.17b}$$

$$\underline{\nu}_j - \underline{\epsilon}_{jk} \leq \theta_{jk}^{\text{RZ}} \leq \bar{\nu}_j + \bar{\epsilon}_{jk}, \quad \forall j \in \tilde{M}, \quad \forall k \in [d+1, d+N] \tag{6.17c}$$

$$\omega(l_k)\underline{u}_j^{\text{irrig}} \leq u_{jk} \leq \omega(l_k)\bar{u}_j^{\text{irrig}}, \quad \forall j \in \tilde{M}, \quad \forall k \in [d, d+N-1] \tag{6.17d}$$

$$l_{\min} \leq l_k \leq l_{\max}, \quad \forall k \in [d, d+N-1] \tag{6.17e}$$

$$\underline{\epsilon}_{jk} \geq 0, \quad \bar{\epsilon}_{jk} \geq 0, \quad \forall j \in \tilde{M}, \quad \forall k \in [d+1, d+N] \tag{6.17f}$$

where $\boldsymbol{l} := [l_d, l_{d+1}, \ldots, l_{d+N-1}]$. NLP solvers can then be employed to efficiently solve the formulation of $\mathbb{P}_{\text{M,N}}^{\text{SIG}}(\boldsymbol{Y}, d)$.

#### 6.3.2.1.1 Determination of the Slope, $\beta$, of $\omega(l)$

Ideally, $\omega(l)$ becomes increasingly similar to binary values (i.e., 0 or 1) as the parameter $\beta$ is increased. Arbitrarily large values of $\beta$, however, may result in numerical instability and a poorly conditioned problem. To address this, an iterative strategy is adopted to progressively enhance the binary approximation of $\omega(l)$ while maintaining numerical stability. This approach involves solving $\mathbb{P}_{\text{M,N}}^{\text{SIG}}(\boldsymbol{Y}, d)$ repeatedly, with $\beta$ scaled by a factor $\tau$ at each iteration, as outlined in Algorithm 2. The process continues until a convergence criterion is met. Specifically, for the $i$th iteration, convergence is achieved when:

$$||\omega(\boldsymbol{l}^i) - \mathbf{c}||_2 \leq \zeta \tag{6.18}$$

where $\mathbf{c}$ is a binary vector obtained by rounding each element of $\omega(\boldsymbol{l}^i)$ to its nearest binary value, and $\zeta$ denotes the prescribed tolerance. The full sequence of steps in this iterative procedure is provided in Algorithm 2.

**Algorithm 2** Iterative Procedure for Approximating $\mathbf{c}$ Using $\omega(\boldsymbol{l})$

**Require:** $\boldsymbol{Y}_d$ (i.e., $\boldsymbol{Y}$ on day $d$), $\boldsymbol{Y}^0$, $\boldsymbol{l}^0$, $(\boldsymbol{U}^{\text{irrig}})^0$, $\zeta$, $\beta^0$, $\tau$

1: $\boldsymbol{Y}_{\text{initial guess}} \leftarrow \boldsymbol{Y}^0$
2: $\boldsymbol{l}_{\text{initial guess}} \leftarrow \boldsymbol{l}^0$
3: $(\boldsymbol{U}^{\text{irrig}})_{\text{guess}} \leftarrow (\boldsymbol{U}^{\text{irrig}})^0$
4: $\beta \leftarrow \beta^0$
5: $i \leftarrow 0$ $||\omega(\boldsymbol{l}^i) - \mathbf{c}||_2 > \zeta$
6: Solve $\mathbb{P}_{\text{M,N}}^{\text{SIG}}(\boldsymbol{Y}_d, d)$ using current $\beta^i$
7: $\boldsymbol{Y}_{\text{initial guess}} \leftarrow \boldsymbol{Y}^i$
8: $\boldsymbol{l}_{\text{initial guess}} \leftarrow \boldsymbol{l}^i$
9: $(\boldsymbol{U}^{\text{irrig}})_{\text{initial guess}} \leftarrow (\boldsymbol{U}^{\text{irrig}})^i$
10: $\beta^{i+1} \leftarrow \tau \cdot \beta^i$
11: $i \leftarrow i + 1$

### 6.3.3 Case Study

A study was carried out on a spatially heterogeneous field segmented into two MZs according to soil texture. One MZ comprised loamy soil, while the other MZ was characterized by loamy sand. Prior to deploying the scheduler $\mathbb{P}^{\text{SIG}}_{\text{M,N}}(\boldsymbol{Y}, d)$, individual LSTM networks were developed for each MZ to improve computational efficiency. The training data for these networks were generated using two separate 1D Richards' equations, each simulating soil moisture dynamics for a respective MZ. These models were calibrated using the hydraulic parameters for loamy and loamy sand soils listed in Table 6.1. Each soil domain was 0.60 m deep and divided into 30 uniformly spaced layers. Pressure head values for all compartments in both MZs were randomly initialized.

In the data generation step, inputs such as irrigation amount, rainfall, daily reference evapotranspiration, and crop coefficient were randomly sampled from the following bounds: (0.33 cm/day, 3.56 cm/day), (1.04 mm, 7.0 mm), (1.04 mm, 3.0 mm), and (0.50, 0.88), respectively. Open-loop simulations were performed with a time step of 360 seconds. Soil moisture at the 0.50-m depth was selected to depict moisture conditions in the root zone. A dataset comprising 56,000 entries was obtained for each MZ. The surrogate models were developed using the Keras library to forecast one-day-ahead root zone moisture. Each model consisted of two hidden layers, each with 200 LSTM units, and was trained on 5-day sequence length.

The limits of the target soil moisture zone were set at 0.256 $m^3/m^3$ (lower limit) and 0.274 $m^3/m^3$ (upper limit) for the loamy soil MZ, and 0.11 $m^3/m^3$ (lower limit) and 0.19 $m^3/m^3$ (upper limit) for the loamy sand MZ. Relevant weights for the loamy MZ were fixed as $\bar{Q} = 9,000$ and $\underline{Q} = 9,000$, while those for the loamy sand were specified as $\bar{Q} = 2,000$ and $\underline{Q} = 9,000$. The per-unit cost of initiating an irrigation event was fixed as $R_c = 50$, and the per-unit water application was set as $R_u = 20$ for both zones. The scheduler was evaluated with over a 14-day horizon. Evapotranspiration values for the simulation period are illustrated in Figure A.2 in Appendix A.2, and a constant crop coefficient of 0.5 was applied. The initial soil moisture content was set to 0.266 $m^3/m^3$ for the loam MZ and 0.127 $m^3/m^3$ for the loamy sand MZ.

Two 1D Richards' equation models—each corresponding to the loam and loamy sand MZs, respectively—served as the simulation environment for the actual field. Root zone soil moisture levels used to update the scheduler were assumed to be known and corresponded to the moisture content at a 0.5-m depth. The scheduler was evaluated over a 20-day simulation window, with optimization performed daily using the interior point optimization (IPOPT) solver [74]. For the implementation of Algorithm 2, the following settings were applied: the slope update factor $\tau = 2.5$, the convergence tolerance $\zeta = 0.01$, and slope bounds defined as $l_{\text{min}} = -5$ and $l_{\text{max}} = 5$. The initial slope value was set to 10.

#### 6.3.3.1 Results

The application of Algorithm 2 revealed that a slope value of 25 effectively promoted convergence of the sigmoid function toward binary outputs while avoiding numerical issues. The irrigation prescriptions and associated root zone soil moisture trajectories for both the loam and loamy sand MZs are presented in Figure 6.5. As illustrated in the figure, the

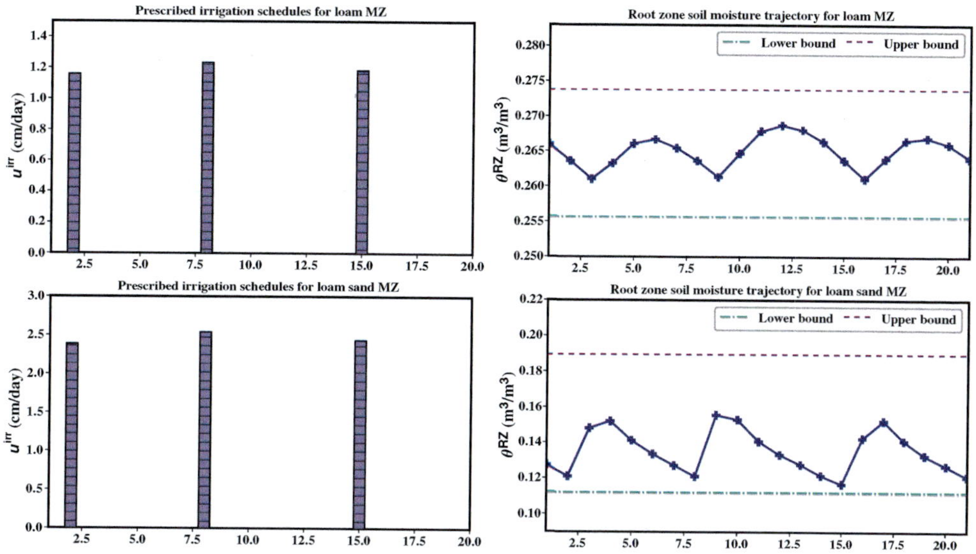


**Figure 6.5** Prescribed irrigation and corresponding root zone soil moisture trajectories for the spatially variable field.

scheduler initiates irrigation events on days 2, 9, and 16 for both zones. In the loam MZ, irrigation depths of 1.20, 1.30, and 1.25 cm/day are applied to maintain soil moisture within the specified target band. In contrast, the loamy sand MZ requires larger applications—2.40, 2.50, and 2.55 cm/day—due to its lower water retention capacity, in order to meet the zone-specific soil moisture targets.

## 6.4 Conclusion

This chapter presented a mixed-integer MPC framework for daily irrigation scheduling, designed to optimize both the timing and volume of irrigation across homogeneous and spatially variable agricultural fields. The scheduler seeks to maintain soil moisture within prescribed bounds while minimizing water usage and associated costs. In uniform field settings, LSTM networks were employed to emulate soil moisture dynamics, significantly reducing the computational burden without sacrificing predictive accuracy. To address spatial variability, fields were subdivided into multiple MZ, each treated as internally homogeneous. A separate LSTM model was trained for each zone to capture local moisture behavior. These zone-specific surrogates were integrated into the MPC formulation to compute targeted irrigation schedules across the field. The performance of the proposed framework was validated through simulation studies. Results demonstrated that the LSTM models provided reliable forecasts of soil moisture, enabling the MPC scheduler to generate effective irrigation plans. For heterogeneous fields, binary irrigation decisions were approximated using a sigmoid-based relaxation, allowing the problem to be recast as a nonlinear program.

This approach retained fidelity to the binary outputs while improving computational efficiency through a careful tuning of the sigmoid function's slope. By synchronizing irrigation timing across zones while allowing zone-specific irrigation amounts, the scheduler maintained balanced moisture levels throughout the field. Overall, the integration of surrogate modeling with a mixed-integer MPC framework offers a scalable solution for precision irrigation, with benefits for resource efficiency and sustainable crop management.

# 7

# Multi-agent MPC for Irrigation Scheduling: A Learning-based Approach

## 7.1 Introduction

Recent advances in computational capabilities, combined with the growing availability of agricultural data, have made it possible for machine learning to significantly enhance the accuracy of irrigation scheduling. Model predictive control (MPC) has similarly gained traction as a key tool for optimizing when and how much to irrigate, ensuring that crops receive water in the right amount at the right time. Collectively, machine learning and MPC now serve as key components in the development of closed-loop irrigation systems that are both adaptive and resource-efficient.

Chapter 6 introduced management zones (MZs) as a practical method for handling spatial heterogeneity in agricultural fields. This approach enables site-specific irrigation by tailoring water application to the distinct needs of different field sections. However, the case study in Chapter 6 relied on a predetermined set of MZs, without tackling the challenge of defining these zones in real-world scenarios. Delineating effective MZs in agricultural fields remains a complex task. Unsupervised learning methods are frequently used to define MZs based on spatial datasets like yield maps, elevation, and soil electrical conductivity. While these techniques can reveal useful patterns, they often overlook key variables like soil hydraulic properties due to the practical difficulty of collecting such data at field-scale. However, as discussed in Chapter 4, soil hydraulic parameters can be estimated from field measurements, offering a more robust foundation for MZ delineation. Incorporating these parameters allows for the development of zones that are not only spatially meaningful but also aligned with model-based irrigation scheduling approaches.

Chapter 6 highlighted the need for soil moisture models that are both accurate and efficient enough to be used within MPC frameworks for irrigation scheduling. Supervised learning has been widely applied for predicting root zone soil moisture. To improve spatial resolution across the soil profile, these models are sometimes used in ensemble setups, where multiple predictions are combined using weighting techniques. This allows for the estimation of moisture distribution along the rooting depth, which is essential for making informed irrigation decisions. However, while ensemble approaches offer better accuracy, they also increase computational requirements—posing challenges when embedded

*Precision Irrigation for Agriculture: Integrating Machine Learning and Optimal Control Strategies*, First Edition. Bernard Twum Agyeman and Jinfeng Liu.

within optimization-based approaches. A further limitation of existing supervised learning models is their assumption of a fixed rooting depth. In reality, root systems expand over time, affecting how water is absorbed at different depths. Current approaches rarely account for this variability, which can lead to inaccurate predictions under changing crop conditions. This gap highlights the need for learning models that can adapt to dynamic rooting depth while remaining computationally suitable for MPC applications.

While Chapter 6 demonstrated the potential of mixed-integer MPC for daily irrigation scheduling in both uniform and spatially variable fields, it also highlighted a key limitation: the significant computational burden associated with solving mixed-integer problems, particularly when multiple MZs are involved. This challenge is further amplified by the limited computational resources typically available at the farm level, making real-time schedule generation difficult in practice. To address this, Chapter 6 explored the use of the logistic sigmoid function as a continuous relaxation to approximate binary irrigation decisions. Although this approach helped reduce computational complexity, it introduced issues such as numerical instability and approximation errors, which in turn affected the accuracy of the irrigation schedules. These limitations point to the need for alternative strategies that are both computationally efficient and accurate in real-world settings. Reinforcement learning (RL), with its capacity to handle complex, sequential decision-making problems, emerges as a promising alternative.

RL has conventionally been used in irrigation systems to handle environmental uncertainties, optimize irrigation timing and quantities, and automate data collection from sensors [55, 57, 102]. Innovations within the field of RL, such as hybrid actor-critic algorithms [58] and hierarchical frameworks [59, 103], have shown promise in handling both discrete and continuous decision variables. This makes them well-suited for addressing mixed-integer optimal control problems. Consequently, for daily irrigation scheduling problems formulated as mixed-integer optimal control problems, RL-based approaches have the potential to be more computationally efficient than MPC-based methods. Furthermore, multi-agent reinforcement learning (MARL), which involves a group of collaborating RL agents, offers a scalable solution for managing irrigation in spatially diverse fields. By reformulating the daily irrigation scheduling problem in spatially variable settings as a multi-agent decision-making problem, MARL can significantly improve the computational performance of scheduling systems.

Recognizing the individual advantages of supervised learning, unsupervised clustering, RL, and MPC, this chapter presents a unified framework for closed-loop irrigation scheduling that integrates all these methodologies. The chapter begins by detailing a three-stage MZ delineation process that combines $k$-means clustering with estimated soil hydraulic properties and topographical data. It then describes the use of LSTM networks to develop surrogate models that capture root zone soil moisture dynamics for each MZ. Next, it introduces a decentralized MARL framework aimed at improving the computational efficiency of mixed-integer control-based schedulers for heterogeneous fields. Finally, the chapter presents a real-field case study to validate the proposed framework, benchmarking its performance against traditional threshold-based (triggered) irrigation methods. The results presented in this chapter were previously published in [104, 105].

## 7.2 Three-stage Process for Management Zone (MZ) Delineation

Establishing effective MZs is an important initial step in developing closed-loop irrigation schedulers. This section presents a three-stage method for MZ delineation, leveraging $k$-means clustering and estimated hydraulic properties to produce MZs that are well-suited for agro-hydrological modeling. Each stage is described in detail below:

1. **Stage 1: Crop-type-based Delineation:** The first stage divides the agricultural field into areas where the same crop type is cultivated. This step is essential, as different crops have distinct water demands that must be considered when scheduling irrigation. Given the frequent presence of several crops in irrigated fields, this crop-type-based grouping ensures that subsequent delineation stages address the unique water requirements of each crop-type.
2. **Stage 2: Integration of Hydraulic Properties and Elevation:** The crop-type-based zones from Stage 1 are further refined using estimated soil hydraulic parameters and elevation data. Key hydraulic properties—including saturated hydraulic conductivity ($K_s$), saturated and residual moisture contents ($\theta_s$, $\theta_r$), and curve-fitting parameters ($\alpha$, $n$)—are essential for accurately modeling soil moisture dynamics and water availability. Elevation data, which can be obtained through field surveys or digital elevation models, is also included, as it influences water redistribution and retention across the field.

   Soil parameters are estimated using the offline state and parameter estimation approach introduced in Chapter 4. After assembling the hydraulic and elevation data, the features are normalized to account for differences in scale. The optimal number of MZs is then determined using the elbow method, followed by the application of $k$-means clustering to identify spatially distinct zones. Additional background on $k$-means clustering is provided in Section 3.1.2.1 of Chapter 3.
3. **Stage 3: Refinement for Irrigation Equipment Compatibility:** The zones produced in Stage 2 may be irregularly shaped and poorly aligned with the layout and resolution of the irrigation infrastructure. To address this, each zone is spatially refined into smaller, regularly shaped subzones that match the operational resolution of the irrigation system. This adjustment enhances the precision of water application, improves system compatibility, and supports more uniform and efficient irrigation across the field.

## 7.3 LSTM-based Modeling of Soil Moisture

Following the delineation of MZs in Section 7.2, the next phase in developing a model-based irrigation scheduler involves constructing models that can simulate soil moisture dynamics efficiently and accurately within each identified zone. This section introduces a data-driven modeling framework that employs LSTM networks to approximate root zone soil moisture behavior. Unlike the method outlined in Chapter 6, the approach adopted here introduces a more practical representation of root zone moisture content. In addition, the model explicitly accounts for temporal changes in rooting depth, improving its adaptability and relevance for real-world irrigation scenarios.

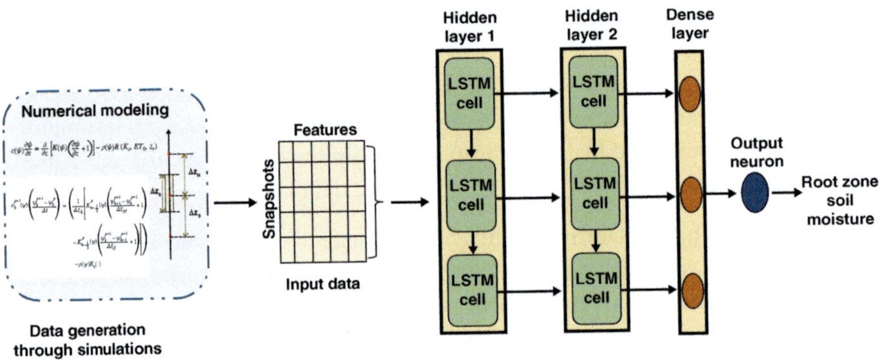


**Figure 7.1** Workflow for LSTM network training. *Source*: [105] with permission of Elsevier/CC BY 4.0.

While long-term field observations of soil moisture and weather conditions are ideal for training such models, they are often unavailable or incomplete. In such cases, synthetic datasets generated from physically based soil water flow models offer a reliable alternative. In this work, open-loop simulations from a calibrated Richards' equation model—illustrated in Figure 7.1—are used to generate training data for each MZ.

The MZ delineation process incorporates soil hydraulic parameters obtained through calibration of a mechanistic model using field sensor data. As a result, the centroidal parameters of each zone are considered representative of the dominant soil hydraulic behavior within that zone. These parameters are used to calibrate a zone-specific mechanistic model based on the Richards' equation, enabling the simulation of soil moisture responses under various irrigation and weather conditions. The simulated data are then used to train LSTM networks tailored to each MZ.

This section first outlines the setup of the physical soil model using the Richards' equation and then describes the LSTM training process. The resulting surrogate models are designed to provide accurate estimates of root zone moisture content while dynamically adjusting to changes in rooting depth.

### 7.3.1 Richards' Equation

As in Chapter 6, soil moisture dynamics within each MZ are modeled using the 1D form of the Richards' equation. This modeling approach is appropriate given that elevation data were explicitly included during the delineation of MZs. Consequently, each zone is expected to exhibit minimal elevation variation, making the 1D formulation a suitable and efficient representation. The 1D Richards' equation, first introduced in Chapter 2, is given as follows:

$$C(\psi)\frac{\partial \psi}{\partial t} = \frac{\partial}{\partial z}\left[K(\psi)\left(\frac{\partial \psi}{\partial z}+1\right)\right] - \rho(\psi)\mathcal{R}\left(K_c, \mathrm{ET}_0, z_r\right) \tag{7.1}$$

To numerically solve the 1D Richards' equation, a set of boundary conditions must be specified. The conditions applied in this chapter are as follows:

$$\left.\frac{\partial(\psi+z)}{\partial z}\right|_{z=H_Z} = 1 \tag{7.2}$$

$$\left.\frac{\partial \psi}{\partial z}\right|_{z=0} = -1 - \frac{u^{\mathrm{irrig}} - \mathrm{EV}}{K(\psi)} \tag{7.3}$$

To calibrate the 1D Richards' equation, the hydraulic properties associated with the centroid of each MZ are used to define the key functions: $K(\cdot)$ (the reader is referred to Equation 2.22), $C(\cdot)$ (the reader is referred to Equation 2.23), and $\theta_v(\cdot)$ (the reader is referred to Equation 2.9). Once these functions are defined, the Richards' equation is solved numerically using the approach outlined in Chapter 2, resulting in the following state-space representation:

$$x_{k+1} = \mathcal{F}(x_k, u_k, \hat{\theta}) + \omega_k \tag{7.4}$$

$$y_k = \mathcal{H}(x_k, \hat{\theta}) \tag{7.5}$$

where $x_k \in \mathbb{R}^{N_x}$ defines the system state, comprising $N_x$ values of capillary pressure head distributed across spatial points within the soil profile. The input $u_k$ captures relevant weather and management variables, including rainfall, irrigation amount, crop coefficient, rooting depth, and reference evapotranspiration. Model uncertainty is represented by $\omega_k$, and the hydraulic behavior of the soil is characterized by $\hat{\theta}$, which contains five parameters evaluated at the centroid of each MZ. The model output is the volumetric water content, $\theta_v$, expressed as $y_k$. This output, $y_k \in \mathbb{R}^{N_y}$, contains $N_p = N_x$ values representing the soil moisture content at the spatial points.

## 7.3.2 Training Data Generation

Equations (7.4) and (7.5) are solved under varying conditions to construct the dataset used for training. Each run starts from a randomly chosen initial state $x_0$ and is driven by a unique input sequence $u_k$. To reflect modeling uncertainty and improve the generalization of the LSTM model, random noise is added during the simulations. This process yields a dataset that captures a wide range of dynamics across different scenarios.

Effective irrigation scheduling requires a reliable estimate of soil moisture distribution across the entire root zone of each MZ. After obtaining the time series data, the output vector $y_k$, which captures moisture content at various soil depths, is used to compute the root zone water content at each time step. To derive a single representative value, the rooting depth $z_r$ is partitioned into four equal layers. The average moisture content is calculated within each layer, and these values are then combined using a weighted scheme: 40% for the top quarter, 30% for the second, 20% for the third, and 10% for the lowest quarter. This weighted average offers an informative summary of moisture conditions across the soil profile, making it suitable for use in irrigation scheduling.

## 7.3.3 Model Design

To support daily irrigation decisions, the LSTM network is designed to predict root zone moisture one day in advance. Specifically, the model forecasts the soil moisture content at time $t+1$, $\hat{y}_{t+1}$, using a sequence of recent input data spanning the previous $l+1$ time steps (i.e., from $t-l$ through $t$, where $l$ is the input history length, which is also known as the sequence length).

Each input vector $\gamma$ supplied to the network consists of five features: the predicted root zone moisture content ($\hat{y}$), the crop coefficient ($K_c$), reference evapotranspiration ($ET_0$), total precipitation (rainfall plus irrigation, $u^{\text{irrig}}$), and rooting depth ($Z_r$). The inclusion of rooting depth enables the model to account for temporal variation in crop development and root expansion, both of which affect moisture availability and uptake. The predictive model is mathematically denoted as:

$$\hat{y}_{t+1} = \mathcal{F}(\{\gamma\}_{t-l}^{t}, \eta) \tag{7.6}$$

where $\{\gamma\}_{t-l}^{t} = [\gamma_{t-l}, \gamma_{t-l+1}, \ldots, \gamma_t]$ denotes the sequence of input vectors, and $\eta$ represents the full set of trainable parameters (i.e., weights and biases) of the LSTM function $\mathcal{F}$.

## 7.4 Mixed-integer MPC with Zone Control for Irrigation Scheduling

Although Chapter 6 offered a comprehensive discussion of the mixed-integer MPC framework with zone control for irrigation scheduling in both homogeneous and heterogeneous fields, the main formulations are restated here to provide the necessary context for subsequent sections.

For a given day $d$, marking the start of the scheduling horizon of length $N$, and an initial soil moisture profile $y$, the mixed-integer MPC-based scheduling problem for a homogeneous field—or equivalently, a single MZ—is denoted by $\mathbb{P}_{\mathrm{N}}(y, d)$ and is defined as follows:

$$\min_{\bar{\epsilon}, \underline{\epsilon}, \boldsymbol{u}^{\mathrm{irrig}}, \boldsymbol{c}} \sum_{k=d+1}^{d+N} \left[\bar{Q}\bar{\epsilon}_k^2 + \underline{Q}\underline{\epsilon}_k^2\right] + \sum_{k=d}^{d+N-1} R_c c_k + \sum_{k=d}^{d+N-1} R_u u_k^{\mathrm{irrig}} \tag{7.7a}$$

s.t.

$$y_{k+1} = \mathcal{F}(\{\gamma\}_{k-l}^{k}, \eta) \qquad k \in [d, d+N-1] \tag{7.7b}$$

$$y_d = y(d) \tag{7.7c}$$

$$\underline{\nu} - \underline{\epsilon}_k \leq y_k \leq \bar{\nu} + \bar{\epsilon}_k, \qquad k \in [d+1, d+N] \tag{7.7d}$$

$$c_k \underline{u}^{\mathrm{irrig}} \leq u_k^{\mathrm{irrig}} \leq c_k \bar{u}^{\mathrm{irrig}}, \qquad k \in [d, d+N-1] \tag{7.7e}$$

$$c_k = \{0, 1\}, \qquad k \in [d, d+N-1] \tag{7.7f}$$

$$\underline{\epsilon}_k \geq 0, \quad \bar{\epsilon}_k \geq 0, \qquad k \in [d+1, d+N] \tag{7.7g}$$

Within $\mathbb{P}_{\mathrm{N}}(y, d)$, Equation (7.7b) represents the LSTM-based surrogate model that describes the soil moisture dynamics of the homogeneous field. This network is constructed using the procedure detailed in Section 7.3. Although Chapter 6 introduced the broader structure of the scheduling framework, it did not elaborate on how to determine the target zone bounds, $(\underline{\nu}, \bar{\nu})$. This section fills that gap by outlining a practical approach for computing these bounds. Specifically, within the scheduler, the values of $(\underline{\nu}, \bar{\nu})$ are determined as follows:

$$\bar{\nu} = \theta_{\mathrm{fc}} \tag{7.8}$$

$$\underline{\nu} = \theta_{\mathrm{th}} = \theta_{\mathrm{fc}} - [\mathrm{MAD} \times (\theta_{\mathrm{fc}} - \theta_{\mathrm{wp}})] \tag{7.9}$$

where $\theta_{\mathrm{fc}}$ and $\theta_{\mathrm{wp}}$ denote the upper and lower limits of plant-available water in the soil, corresponding to field capacity and the wilting point, respectively. The management allowable depletion (MAD) defines the threshold fraction of this available water that can be utilized before irrigation becomes necessary to avoid crop stress.

The subsequent formulation presents the mixed-integer MPC-based scheduler designed for spatially heterogeneous fields, which forms the core focus of this chapter. Let $d$ denote the first day of the scheduling horizon of length $N$, and let $\boldsymbol{Y}$ represent the initial soil moisture profile across all $M$ MZs. The corresponding optimization problem is denoted as $\mathbb{P}_{\mathrm{M,N}}(\boldsymbol{Y}, d)$ and is defined as follows:

$$\min_{\bar{\boldsymbol{E}}, \underline{\boldsymbol{E}}, \boldsymbol{U}^{\mathrm{irrig}}, \boldsymbol{c}} \sum_{j=1}^{M} \sum_{k=d+1}^{d+N} \left[ \bar{Q}(j)\bar{\epsilon}_{jk}^2 + \underline{Q}(j)\underline{\epsilon}_{jk}^2 \right] + R_c \sum_{k=d}^{d+N-1} c_k + R_u \sum_{j=1}^{M} \sum_{k=d}^{d+N-1} u_{jk}^{\mathrm{irrig}} \tag{7.10a}$$

s.t.

$$y_{j(k+1)} = \mathcal{F}^j(\{\gamma_j\}_{k-l}^{k}, \eta_j), \ j \in \tilde{M}, \qquad k \in [d, d+N-1] \tag{7.10b}$$

$$y_{j0} = y_j(0), \qquad j \in \tilde{M} \tag{7.10c}$$

$$\underline{\nu}_j - \underline{\epsilon}_{jk} \leq y_{jk} \leq \bar{\nu}_j + \bar{\epsilon}_{jk}, \ j \in \tilde{M}, \qquad k \in [d+1, d+N] \tag{7.10d}$$

$$c_k \underline{u}_j^{\mathrm{irrig}} \leq u_{jk} \leq c_k \bar{u}_j^{\mathrm{irrig}}, \quad j \in \tilde{M}, \qquad k \in [d, d+N-1] \tag{7.10e}$$

$$c_k \in \{0, 1\}, \qquad k \in [d, d+N-1] \tag{7.10f}$$

$$\underline{\epsilon}_{jk} \geq 0, \quad \bar{\epsilon}_{jk} \geq 0, \qquad j \in \tilde{M}, \qquad k \in [d+1, d+N] \tag{7.10g}$$

The formulation $\mathbb{P}_{\mathrm{M,N}}(\boldsymbol{Y}, d)$ extends its single-zone counterpart, $\mathbb{P}_{\mathrm{N}}(y, d)$, to a multi-zone setting by applying the scheduling problem independently to each of the $M$ MZs within the field. Although each zone is treated separately, a coupling constraint—introduced in Equation (7.10e)—ensures uniform irrigation timing decisions across all MZs throughout the scheduling horizon. A comprehensive discussion of both formulations is provided in Chapter 6.

## 7.5 Decentralized Hybrid Actor-critic Agents and the Role of a Limiting MZ

Solving $\mathbb{P}_{\mathrm{M,N}}(\boldsymbol{Y}, d)$ using standard mixed-integer optimization methods—such as branch-and-bound—can be computationally intensive, particularly in settings involving several MZs. This computational burden presents a significant obstacle in practical agricultural environments, where on-site computational resources are often limited.

To address these limitations, this chapter presents a decentralized learning framework in which hybrid actor-critic RL agents are trained to make irrigation decisions independently at the level of individual MZs. To ensure coherent field-wide scheduling despite this decentralization, the framework also introduces the notion of a limiting MZ, which provides a common reference for synchronizing irrigation timing decisions across MZs in a computationally efficient manner.

The section begins by detailing the hybrid actor-critic architecture used to train MZ-specific agents. It then describes the mechanism by which these agents operate in parallel and how the limiting MZ is leveraged to coordinate decisions.

### 7.5.1 Agent Design and Training

In the proposed framework, an RL agent is assigned to each of the $M$ MZs distributed across the field. The role of each agent is to determine daily irrigation decisions consisting of a discrete component (i.e., whether irrigation should occur) and a continuous component (i.e., the irrigation amount). These decisions are made with the goal of maintaining root zone soil moisture within a target range while minimizing overall water usage and fixed irrigation costs.

To effectively address this mixed action space, the hybrid actor-critic architecture introduced by [58] is adopted. This architecture is particularly suited to control problems involving both discrete and continuous actions. It features two parallel actor networks: one responsible for learning discrete irrigation decisions and the other for learning continuous irrigation amounts. Both actors are supported by a shared critic network, which evaluates the value of a given state and provides feedback to guide the learning process. Each agent operates within the framework of a fully observable Markov decision process (MDP), defined by its observation space, action space, reward function, and the transition dynamics of the underlying environment.

This section proceeds as follows. First, it describes the simulated training environment used to develop the RL agents. Then, it outlines the structure of the observation space—the set of input features informing each agent's policy. The reward function, which encodes the objectives guiding agent behavior, is then detailed. Lastly, the training methodology used to optimize the policy networks is presented.

#### 7.5.1.1 Environment Design

The training environment for each agent is defined by the soil moisture dynamics modeled using the 1D Richards' equation, as specified in Equations (7.4) and (7.5). These dynamics are parameterized using the hydraulic properties associated with the centroid of the respective MZ, thereby ensuring that the simulated environment accurately represents the soil behavior characteristic of that zone.

#### 7.5.1.2 Observation Design

At each step $k$, the agent's policy receives an observation vector composed of key environmental and crop-related variables. This vector is defined as:

$$o_k^{rl} := [y_k, \mathrm{ET}_{0_k}, \mathrm{K}_{c_k}, \mathrm{z}_{r_k}]$$

where $y_k$ denotes the volumetric soil moisture content derived from Equation (7.5), $\mathrm{ET}_{0_k}$ is the reference evapotranspiration, $\mathrm{K}_{c_k}$ represents the crop coefficient, and $\mathrm{z}_{r_k}$ indicates the rooting depth on day $k$.

Among the components of $o_k^{\mathrm{rl}}$, $y_k$ serves as the primary indicator of moisture availability within the MZ. Its initial value is determined using the converged soil moisture estimates obtained during Stage 2 of the MZ delineation process, ensuring that each agent starts from a soil moisture profile that accurately reflects conditions within its respective zone. The remaining inputs in $o_k^{\mathrm{rl}}$ are also readily available: $\mathrm{ET}_{0_k}$ can be obtained from weather forecast data, $\mathrm{K}_{\mathrm{c}_k}$ is derived from empirical crop models, and $\mathrm{z}_{\mathrm{r}_k}$ can be estimated based on the crop's phenological stage. Given the accessibility of all elements in $o_k^{\mathrm{rl}}$, the environment is treated as fully observable.

#### 7.5.1.3 Reward Design

The reward signal $r_k$, which plays a key role in shaping the agent's behavior, is computed at each training step $k$ based on the following formulation:

$$r_k := -R_k^Z - R_c c_k - R_u u_k^{\mathrm{irrig}} \tag{7.11}$$

$$R_k^Z := \begin{cases} \underline{Q} \times |\theta_k^{\mathrm{RZ}} - \underline{\nu}| & \text{if } \theta_k^{\mathrm{RZ}} < \underline{\nu} \\ \overline{Q} \times |\theta_k^{\mathrm{RZ}} - \overline{\nu}| & \text{if } \theta_k^{\mathrm{RZ}} > \overline{\nu} \\ 0 & \text{if } \underline{\nu} \le \theta_k^{\mathrm{RZ}} \le \overline{\nu} \end{cases} \tag{7.12}$$

The reward function incorporates constants such as $R_c$, $R_u$, $\underline{Q}$, and $\overline{Q}$, along with the moisture bounds $\underline{\nu}$ and $\overline{\nu}$, all defined consistently with the formulation $\mathbb{P}_{\mathrm{N}}(y, d)$. The term $\theta_k^{\mathrm{RZ}}$ is derived through the averaging approach of Section 7.3.2.

The reward structure is designed to penalize deviations from the moisture range. If $\theta_k^{\mathrm{RZ}} < \underline{\nu}$, the agent receives a penalty proportional to the shortfall, computed as $-\underline{Q} \times |\theta_k^{\mathrm{RZ}} - \underline{\nu}|$. Conversely, when $\theta_k^{\mathrm{RZ}} > \overline{\nu}$, a similar penalty of $-\overline{Q} \times |\theta_k^{\mathrm{RZ}} - \overline{\nu}|$ is imposed. A reward of 0 is granted if $\theta_k^{\mathrm{RZ}}$ remains within the target range $[\underline{\nu}, \overline{\nu}]$. Beyond these moisture-related penalties, the agent is further incentivized to avoid excessive irrigation through cost-based penalties. A fixed penalty of $-R_c c_k$ is applied whenever irrigation is activated, and a variable penalty of $-R_u u_k^{\mathrm{irrig}}$ scales with the amount of water applied.

#### 7.5.1.4 Agent Training

The parallel actor networks within the hybrid actor-critic framework are trained using the proximal policy optimization (PPO) algorithm [73], a method described in greater detail in Section 3.1.3.5.2 of Chapter 3.

During training, the discrete actor network learns a stochastic policy for binary irrigation decisions. It outputs two logits corresponding to the possible actions $\{0, 1\}$, from which a decision is sampled according to a softmax distribution. Concurrently, the continuous actor network models the irrigation amount by predicting the mean and variance of a Gaussian distribution, enabling the sampling of continuous irrigation amounts. A schematic of this hybrid PPO architecture, as implemented for a single MZ or a homogeneous field, is illustrated in Figure 7.2.

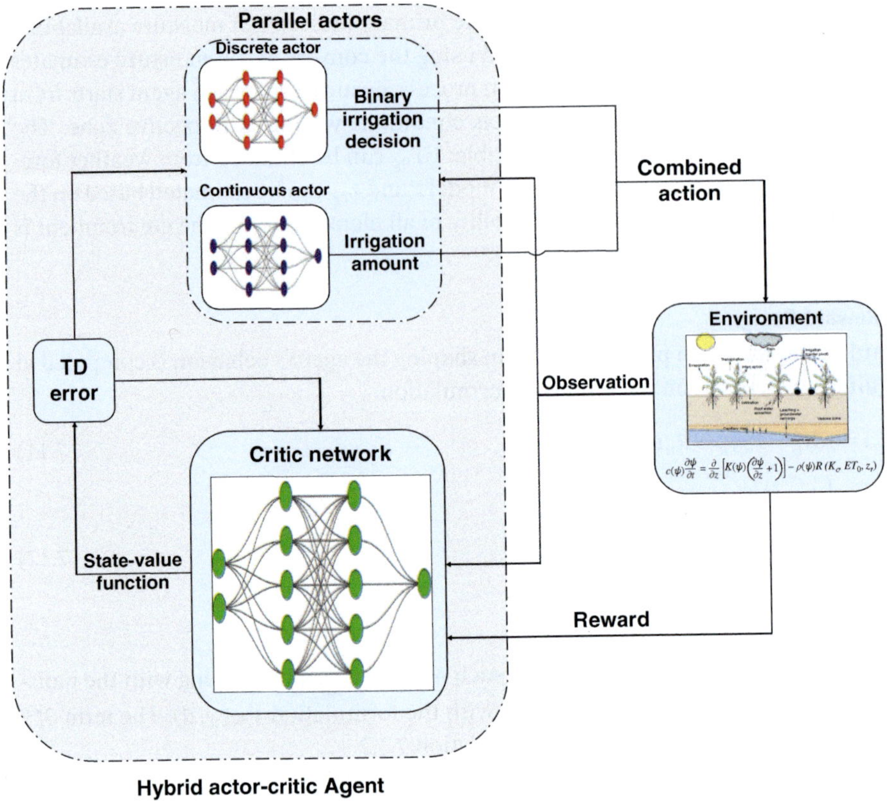


**Figure 7.2** Architecture of the hybrid actor-critic agent and its interaction with the environment. *Source*: [105] with permission of Elsevier/CC BY 4.0.

## 7.5.2 Multi-agent MPC Paradigm

Once training is complete, each of the $M$ agents is deployed to generate irrigation schedules for its corresponding MZ. These schedules consist of binary irrigation decisions and their associated irrigation amounts, computed over $N$. In this evaluation, each agent receives the initial soil moisture content on day $d$, along with a sequence of forecasted uncontrolled inputs—including reference evapotranspiration, crop coefficient, and rooting depth—for the upcoming $N$ days. Using this information, the agent simulates its interactions with a calibrated 1D Richards' equation model tailored to its specific MZ. This process is illustrated in Figure 7.3, where $y_d$ represents the initial soil moisture state and $m_{d+N-1}$ denotes the vector of predicted uncontrolled inputs over the horizon. At each time step, the agent outputs a binary irrigation decision $c$ and a corresponding irrigation amount $u^{\text{irrig}}$. Over the full horizon, this results in two output sequences: $[c_d, \ldots, c_{d+N-1}]$ and $\left[u_d^{\text{irrig}}, \ldots, u_{d+N-1}^{\text{irrig}}\right]$. The process is independently executed by all $M$ agents, producing $M$ binary decision sequences and $M$ irrigation amount sequences.

The independent training of the $M$ hybrid agents results in each agent producing a distinct binary irrigation decision sequence specific to its assigned MZ. This lack of synchronization leads to inconsistencies that are incompatible with the structure of $\mathbb{P}_{\text{M,N}}(\boldsymbol{Y}, d)$, which

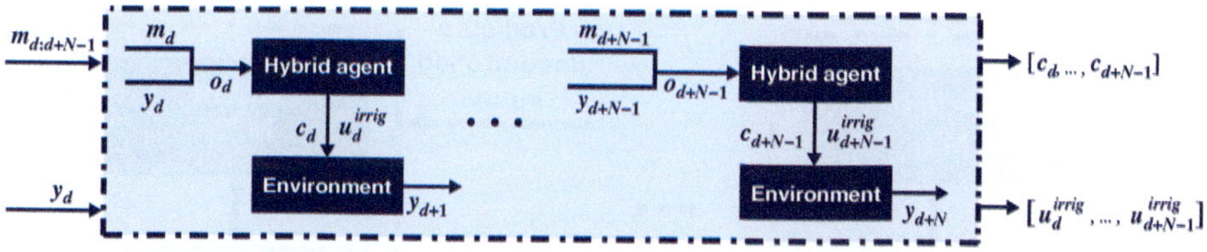


**Figure 7.3** Workflow of the interaction between an agent and its MZ. The process starts with the initial soil moisture state $y_d$ and the sequence of predicted uncontrolled inputs $m_{d:d+N-1}$. On each day $d$, the agent receives the observation $o_d := [y_d, m_d]$, where $m_d$ is the uncontrolled input. Based on this input, the agent outputs $c_d$ and $u_d^{\text{irrig}}$. These actions are applied to the environment, resulting in a transition to the next soil moisture state $y_{d+1}$. This cycle continues until $d+N-1$, producing the sequences $[c_d, \ldots, c_{d+N-1}]$ and $\left[u_d^{\text{irrig}}, \ldots, u_{d+N-1}^{\text{irrig}}\right]$. *Source*: [105] with permission of Elsevier/CC BY 4.0.

requires a unified binary irrigation sequence to be applied across all $M$ zones. To resolve this issue, the framework adopts a limiting MZ selection strategy, wherein the decision sequence from a single representative MZ is used to coordinate irrigation timing throughout the field.

The $jth$ MZ whose binary decision sequence includes the earliest occurrence of a nonzero binary decision is designated as the limiting MZ. Once identified, the binary sequence from this zone is designated as the binding irrigation schedule and is uniformly enforced across all $M$ zones. This selection ensures that irrigation is triggered in a timely manner, guided by the zone with the most immediate water requirement. To demonstrate this concept, consider a field partitioned into three MZs—$\text{MZ}_1$, $\text{MZ}_2$, and $\text{MZ}_3$—each managed by its own agent. Suppose the agents are implemented over a seven-day horizon and generate the following binary irrigation sequences: $\mathbf{c}_{\text{MZ}_1} = [0,1,0,1,0,0,0]$, $\mathbf{c}_{\text{MZ}_2} = [0,0,0,1,0,0,1]$, and $\mathbf{c}_{\text{MZ}_3} = [0,0,0,0,1,0,1]$. In this case, $\text{MZ}_1$ is the first to trigger irrigation (on day 2) and is therefore selected as the limiting MZ. Thus, $\mathbf{c}_{\text{MZ}_1}$, is adopted as the binding irrigation schedule for all zones.

After the selection of the binding irrigation decision sequence, it is applied directly within the formulation $\mathbb{P}_{\text{M,N}}(\boldsymbol{Y}, d)$. With the binary decision variables fixed over $N$, the original mixed-integer formulation becomes significantly simplified. In particular, $\mathbb{P}_{\text{M,N}}(\boldsymbol{Y}, d)$ is transformed from a mixed-integer nonlinear program (MINLP) into a nonlinear program (NLP), enabling more efficient computation of the irrigation amounts required for all $M$ MZs.

The computational burden associated with solving the NLP version of $\mathbb{P}_{\text{M,N}}(\boldsymbol{Y}, d)$ can be substantially reduced, or even eliminated, by leveraging problem structure. Once the binding binary irrigation decision sequence is fixed, the multi-MZ formulation naturally decomposes into $M$ independent NLPs, each corresponding to an instance of $\mathbb{P}_{\text{N}}(y, d)$ for a specific MZ. This decomposition is made possible because the binary decision variable is the sole coupling element across MZs. Aside from this, $\mathbb{P}_{\text{M,N}}(\boldsymbol{Y}, d)$ is essentially a replication of the single-zone problem across the $M$ zones. By resolving the binary variables through the limiting MZ strategy, each zone-specific NLP can be solved independently. These subproblems are amenable to parallel computing techniques, allowing for simultaneous execution on multicore systems and thereby significantly reducing overall computation time. Figure 7.4 summarizes this workflow, illustrating how the trained agents, limiting MZ selection, and parallelized NLP subproblems are integrated within the framework.

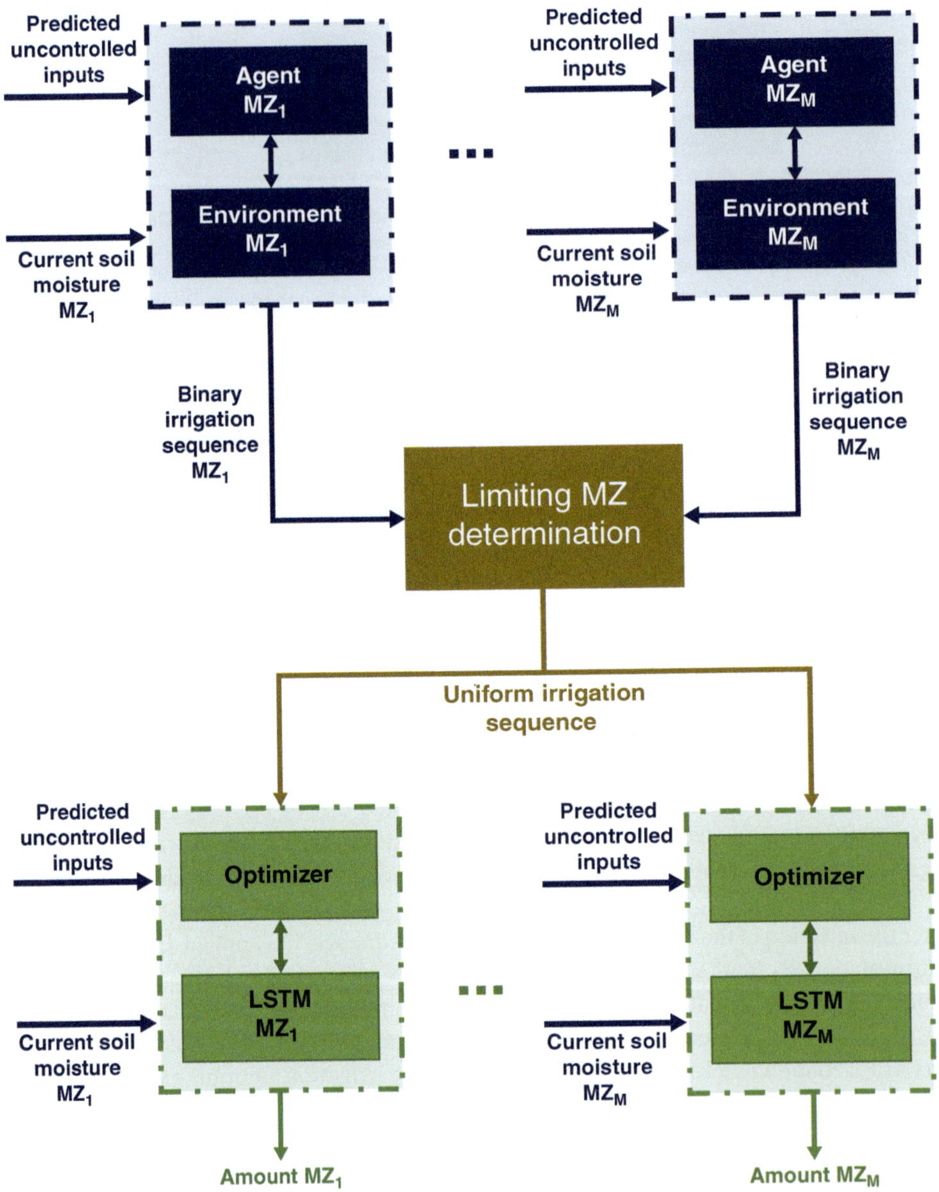


**Figure 7.4** An illustration of the multi-agent MPC paradigm. In this paradigm, individual irrigation decision sequences produced by $M$ hybrid PPO agents are aggregated to form a binding irrigation decision sequence. This binding sequence is then used to solve $M$ separate instances of $\mathbb{P}_N(y, d)$, yielding the irrigation amounts for each MZ. *Source*: [105] with permission of Elsevier/CC BY 4.0.

The prescribed amounts generated by the agent for the limiting MZ can be used without modification, as they are inherently aligned with the fixed binary schedule. A sequence of irrigation amounts is said to be consistent with a binary decision sequence if, on each day of the scheduling horizon, a binary value of 1 corresponds to a nonzero irrigation amount, and a value of 0 corresponds to zero irrigation. In contrast, agents associated with the non-limiting MZs may produce irrigation amounts that violate this consistency. To enforce alignment with the fixed binary sequence, the NLP version of $\mathbb{P}_{\mathrm{N}}(y, d)$ must be solved individually for each of these zones. Although the agent outputs may not initially satisfy the binding sequence, they offer good initial guesses for the NLP solver. When used to warm-start the optimization, these guesses can improve convergence speed.

### 7.5.3 Triggered Irrigation Scheduling

To serve as a benchmark for comparison, a triggered irrigation strategy is employed. This rule-based approach is designed to maintain soil moisture within a specified target interval, bounded below by $\underline{\nu}$ and above by $\overline{\nu}$. An irrigation event is activated on day $d$ whenever $\theta_d^{\mathrm{RZ}} < \underline{\nu}$. If this condition is satisfied, the corresponding irrigation amount $u_d^{\mathrm{irrig}}$ is computed using the equation proposed by [19]:

$$u_d^{\mathrm{irrig}} := \left[\left(\bar{\nu} - \theta_d^{\mathrm{RZ}}\right) \times z_r\right] - P_{d+4} \tag{7.13}$$

where $P_{d+4}$ denotes the cumulative rainfall forecast over the five-day period spanning from day $d$ through day $d+4$.

## 7.6 Application to a Large-scale Field

To assess the effectiveness of the proposed scheduler, a large-scale field at the Lethbridge Polytechnic Research Farm in southern Alberta was used. The field spans 26.4 hectares and is located at approximately 49.72° N latitude and 112.80° W longitude. Its topography varies from 886 to 890 m in elevation, and the dominant soil type is clayey loam, interspersed with sand lenses. A map of the field is shown in Figure 4.5.

The field is irrigated using a five-span center pivot system with a total length of 294 m, which distributes water in a circular pattern from a central point. This system is equipped with variable rate irrigation that enables differential water application across the field. Microwave radiometers are installed along the pivot's arm to monitor soil moisture levels of the field. Data from these radiometers were used to estimate spatial variations in soil water content and hydraulic properties, following the procedure outlined in Chapter 4. These estimates formed the basis for implementing the MZ delineation strategy. Data collection for the estimation study was conducted between May 16, 2022, and August 31, 2022.

### 7.6.1 Delineation of MZs in the Study Area

Figure 7.5 presents the results of the MZ delineation process for the study area, which was conducted in three stages. The first stage involved dividing the field into four quadrants (Quadrants 1 through 4), reflecting the typical cropping practice of planting different crops in each quadrant. For brevity, the delineation results for only Quadrants 1 and 3 are shown.

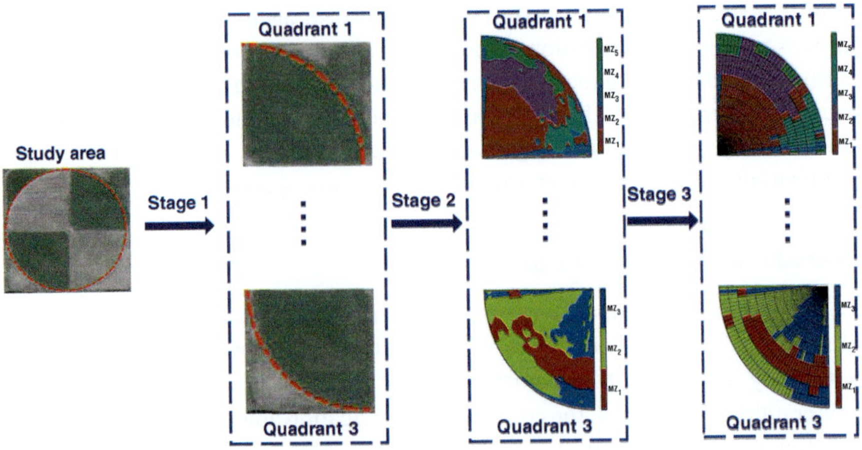


Figure 7.5 Visualization of the outcomes from the three-stage irrigation MZ delineation process as applied to the field. *Source*: [105]/Elsevier/CC BY 4.0.

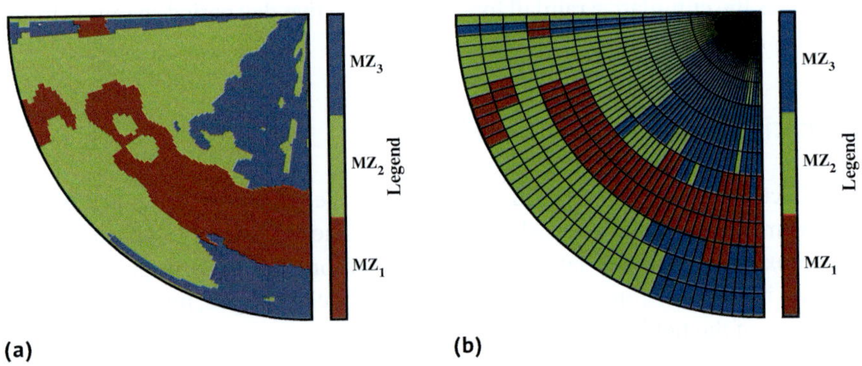


Figure 7.6 MZs and their spatial representation at the resolution of the center pivot system. (a) Delineated MZs exhibiting irregular shapes; (b) MZs represented as pivot resolution. *Source*: [105] with permission of Elsevier/CC BY 4.0.

In stage two, each quadrant was refined further using spatial variations in estimated soil hydraulic properties and elevation data. This process produced five MZs within Quadrant 1 and three within Quadrant 3, as shown in Figure 7.5.

The final stage adjusted the irregular MZs to align with the resolution of the center pivot irrigation system, which operates using 12 radial and 45 azimuthal control zones. For clarity, the analysis hereafter focuses on Quadrant 3. Figure 7.6a displays the three MZs generated in the second stage, while Figure 7.6b shows their resolution-adjusted counterparts. In terms of soil and elevation characteristics, $MZ_1$ in Quadrant 3 consists of loamy soil with an average elevation of 889 m. $MZ_2$ shares the same soil type but is slightly lower at 888 m. $MZ_3$ is characterized by sandy clay loam with an average elevation of 888.5 m.

The remainder of this chapter applies the proposed scheduling framework to Quadrant 3. However, the methodology is general and can be extended to the other quadrants without modification.

### 7.6.2 LSTM Network Training for Quadrant 3

To simulate soil moisture dynamics in Quadrant 3, a 1D Richards' equation was calibrated for each of the three irrigation MZs using their respective centroidal hydraulic parameters. These models were then used to generate training data through open-loop simulations, where both inputs and initial conditions were randomly sampled. The configuration settings for these simulations are provided in Table E.1 of Appendix E.1.

Training of the LSTM networks was conducted using data generated for two rooting depths (0.5 m and 1.0 m) to capture seasonal variability in root development. For each MZ, daily reference evapotranspiration values were sampled uniformly from 0.1 to 8.99 mm/day, crop coefficients from 0.4 to 1.02, and irrigation amounts from MZ—specific ranges: 4.0–52.0 mm/day ($MZ_1$), 4.3–59.6 mm/day ($MZ_2$), and 5.0–62.3 mm/day ($MZ_3$). These ranges reflect local irrigation practices and historical weather patterns. Model architectures and training hyperparameters are summarized in Table E.2 of Appendix E.2.

Model performance was evaluated by comparing LSTM-predicted root zone moisture values to those obtained from the corresponding Richards' equations over a 30-day horizon. Simulations were run with fixed initial conditions and predefined input trajectories, and accuracy was quantified using the Root Mean Squared Error (RMSE) metric.

### 7.6.3 Implementation of Hybrid PPO Agents in Quadrant 3

Three hybrid PPO agents were developed to manage irrigation decisions for the distinct MZs identified in Quadrant 3. The parameters used in $\mathbb{P}_{\mathrm{N}}(y, d)$ and $\mathbb{P}_{\mathrm{M,N}}(\boldsymbol{Y}, d)$, which guided the training of these agents, are summarized in Table E.4 of Appendix E.4. Soil-specific thresholds, including field capacity and wilting point values for each MZ, were obtained from [106], based on the dominant soil textures in the MZs.

The two actor networks in each agent were constructed as two-layer fully connected multilayer perceptrons, with tanh activation functions applied to each hidden layer. Details of the training hyperparameters and PPO algorithm settings are provided in Table E.3 of Appendix E.3.

To assess performance under different irrigation strategies, agents were trained using two levels of MAD: 40% and 65%. For each MAD level, a separate set of three agents—one for each MZ—was trained independently. The training process was carried out in Python with the Tensorforce library [107].

### 7.6.4 Implementation of Proposed and Triggered Schedulers in Quadrant 3

The comparative evaluation of the proposed and triggered scheduling methods in Quadrant 3 was conducted using field data from the 2015 and 2022 seasons. These two seasons were selected to represent distinct climatic conditions, with Figure A.3 of Appendix A.3 illustrating their respective weather profiles. The 2015 season, marked by lower precipitation (141.7 mm), is classified as dry, whereas the 2022 season received a higher total rainfall of 230.9 mm and is considered wet. In both seasons, soft spring wheat was cultivated, and crop coefficient values were derived using Equation (A.3) of Appendix A.4. Two MAD thresholds—40% (conservative) and 65% (less conservative)—were examined to assess performance under varying water stress tolerances.

Soil moisture conditions for each MZ were initialized based on estimates from the delineation method outlined in Section 7.2. The minimum volumetric moisture value identified within each MZ was used to initialize the scheduling algorithm. The proposed model-based scheduler was applied in a receding horizon fashion, with both the prediction and control horizons set to 14 days. To represent actual soil moisture behavior, the calibrated 1D Richards' equations for the three MZs were used.

Although actual weather data were used for evaluation, the forecasted inputs over the scheduling horizon were perturbed to simulate uncertainty. This was achieved by introducing additive noise drawn from a zero-mean normal distribution, with the standard deviation increasing over time to reflect the greater uncertainty of longer-term forecasts.

The configuration parameters used for the formulation $\mathbb{P}_{\mathrm{N}}(y, d)$ are listed in Table E.4 (Appendix E.4). The evaluation spanned 123 days, from May 5 to September 4. At each step in the scheduling horizon, the corresponding NLP was solved using an interior-point optimization algorithm [74]. Only the first element of the resulting irrigation sequence was implemented at each step, with the updated soil moisture state serving as the initial condition for the subsequent optimization. Rooting depth was set at 0.5 m until July 15, after which it was increased to 1.0 m to reflect seasonal root development patterns.

To provide a basis for comparison, a triggered scheduling approach was implemented in parallel. This method evaluated the root zone soil moisture in each MZ daily. If the moisture level in all zones remained above their respective lower bounds, irrigation was skipped. If any zone fell below its threshold, irrigation was applied uniformly across all zones, with the required amounts computed using Equation (7.13).

Performance comparisons between the two methods were based on total irrigation applied and Irrigation Water Use Efficiency (IWUE), also referred to as water productivity. The IWUE, expressed in kilograms per cubic meter (kg/m$^3$), is defined by the following relation:

$$\mathrm{IWUE} = \frac{Y_a}{\mathrm{IRR}} \tag{7.14}$$

where $Y_a$ is the predicted crop yield in kg/m$^2$, and IRR denotes the total irrigation applied (in meters). Details on the yield estimation process are provided in Appendix C.

## 7.6.5 Results and Discussion

The remainder of this section reports and discusses the key findings from deploying the integrated irrigation scheduling framework in Quadrant 3 of the study area. The analysis begins by verifying the predictive ability of the LSTM-based surrogate models. It then reports and discusses the relevant training statistics of the trained agents. The section concludes with a comparison between the model-based scheduler and the triggered rule-based approach.

### 7.6.5.1 Assessment of LSTM Model Accuracy

Figure 7.7 presents the validation results of the trained LSTM models against reference outputs from the calibrated 1D Richards' equation. The close alignment between the LSTM predictions and the reference values highlights the models' effectiveness in capturing soil moisture dynamics. This is further supported by the RMSE values in Table 7.1, which remain low across all MZs and rooting depths of 0.50 and 1.00 m. The models also performed

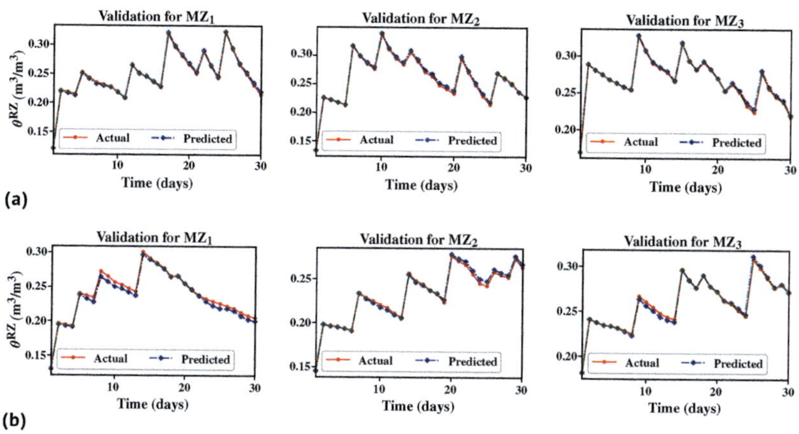


**Figure 7.7** Prediction accuracy of the trained LSTM models across different rooting depths. (a) Rooting depth of 0.50 m; (b) Rooting depth of 1.00 m.

**Table 7.1** Performance of trained LSTM networks in 30-day soil moisture forecasting.

| Rooting depth (m) | MZ | RMSE |
|---|---|---|
| 0.50 | 1 | 0.0053 |
| 1.00 | 1 | 0.0033 |
| 0.50 | 2 | 0.0044 |
| 1.00 | 2 | 0.0048 |
| 0.50 | 3 | 0.0037 |
| 1.00 | 3 | 0.0023 |

well in extended forecasting scenarios. Although initially designed for one-day-ahead predictions, the LSTMs were applied recursively over a 30-day horizon, with each prediction serving as input to the next.

The results demonstrate that the proposed LSTM model provides a reliable means of estimating root zone soil moisture within the MZs of spatially-variable agricultural fields. These findings are consistent with previous research that has used LSTM networks to represent soil moisture behavior in agricultural contexts [22]. Moreover, they support the notion that LSTM networks, by virtue of their universal function approximation capabilities, can estimate root zone moisture directly. This presents a more streamlined and computationally efficient alternative to earlier approaches that employed ensemble models for different soil layers. For instance, prior work [19] developed separate neural networks for various depths and aggregated their outputs to estimate root zone moisture. By contrast, the current model incorporates rooting depth as a dynamic input, enabling it to account for temporal changes in root growth and their influence on soil moisture in a more direct manner.

#### 7.6.5.2 Evaluation of the Trained Hybrid PPO Agents

Figure 7.8 presents the average score trajectories observed during the training of the hybrid PPO agents under the 65% MAD setting. Each curve reflects the mean performance across 10 independent training trials. The upward trends observed in the plots suggest that the agents improved their policies steadily over time. The eventual flattening of the trajectories signals convergence toward stable and effective strategies. These results confirm that the trained agents were able to learn irrigation policies tailored to the specific soil and crop conditions of their corresponding MZs.

In a broader context, this hybrid PPO approach builds on existing strategies aimed at reducing the complexity of mixed-integer control problems. As introduced in Chapter 6, the sigmoid-based method approximates binary decisions using continuous variables, simplifying the optimization process. However, this can lead to approximation errors and reduce the interpretability of the resulting decisions. The hybrid PPO method addresses both limitations by learning binary decisions directly from interaction with the environment, eliminating the need for continuous relaxations. Another commonly used strategy is time-step grouping [108], which simplifies the problem by combining multiple binary decisions across several time steps into a single variable. While this reduces the number of decision

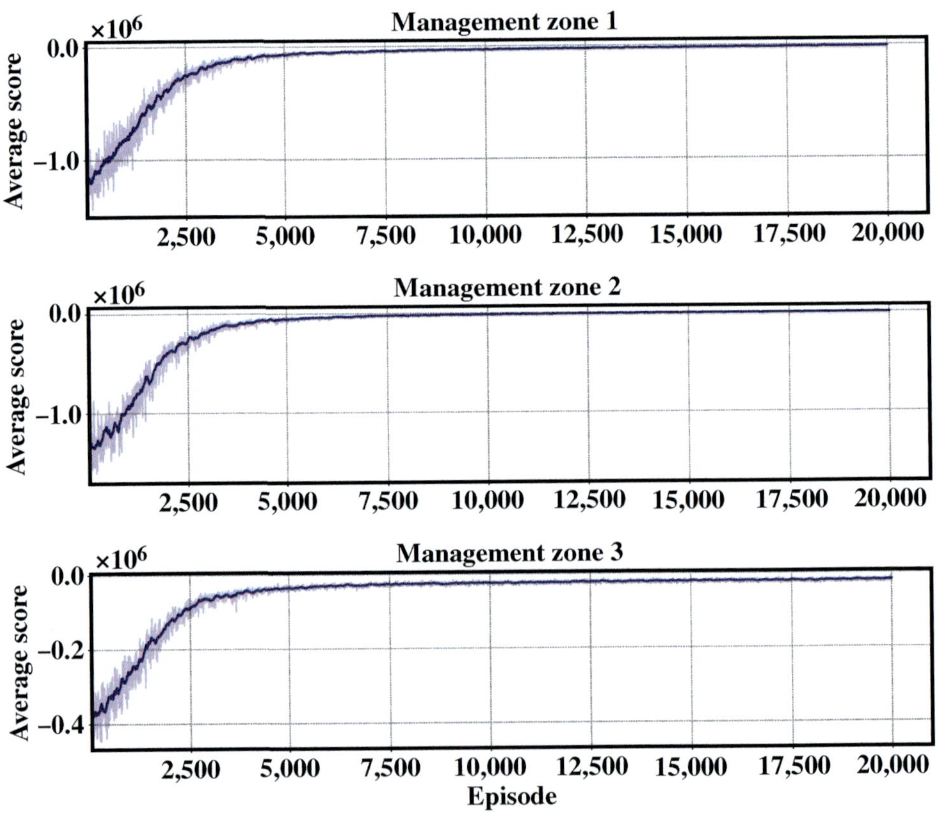


**Figure 7.8** Average score trajectories of the reinforcement learning agents across MZs.

variables, it also requires manual grouping choices and may limit the system's flexibility. In contrast, the hybrid PPO method determines binary actions independently at each time step, offering greater flexibility.

### 7.6.6 Evaluation of the Proposed Scheduler's Effectiveness

Figures 7.9 and 7.10 show the irrigation schedules and the soil moisture trajectories for the 2022 growing season under MAD levels of 40% and 65%, respectively. The corresponding results for the 2015 season are shown in Figures 7.11 and 7.12. Performance metrics are provided in Tables 7.2 and 7.3 for 2022, and Tables 7.4 and 7.5 for 2015.

As expected, lower rainfall in 2015 led to increased irrigation demand across both strategies. In both years, the use of a conservative MAD value (40%) resulted in greater total irrigation and more frequent applications compared to the 65% setting. This outcome is consistent with the rationale behind lower MAD thresholds, which permit less soil moisture depletion before triggering irrigation.

In all scenarios, the proposed scheduling method delivered consistent reductions in water use relative to the triggered approach. For the 40% MAD level, irrigation was reduced by approximately 7% in 2015 and 6% in 2022. With a 65% MAD, savings rose to 14% and 23%, respectively. The proposed method also improved IWUE, indicating more effective water utilization. IWUE gains under the 40% MAD setting were 10% for 2015 and 12% for 2022. With a 65% MAD, improvements reached 20% and 35%, respectively.

These results demonstrate that the proposed framework can achieve substantial water savings and efficiency gains, particularly when used with less conservative MAD values. While lower MAD thresholds are often favored as a safeguard against crop stress, the findings suggest that more relaxed settings can be adopted without compromising crop yields.

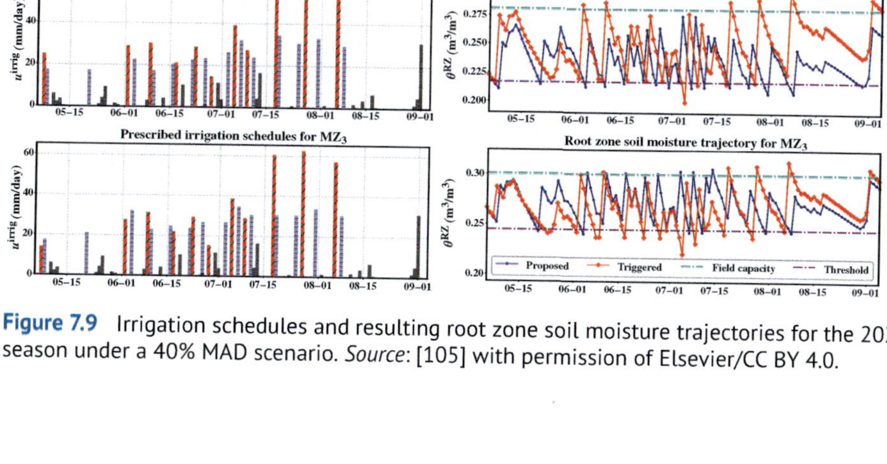


**Figure 7.9** Irrigation schedules and resulting root zone soil moisture trajectories for the 2022 wet season under a 40% MAD scenario. *Source*: [105] with permission of Elsevier/CC BY 4.0.

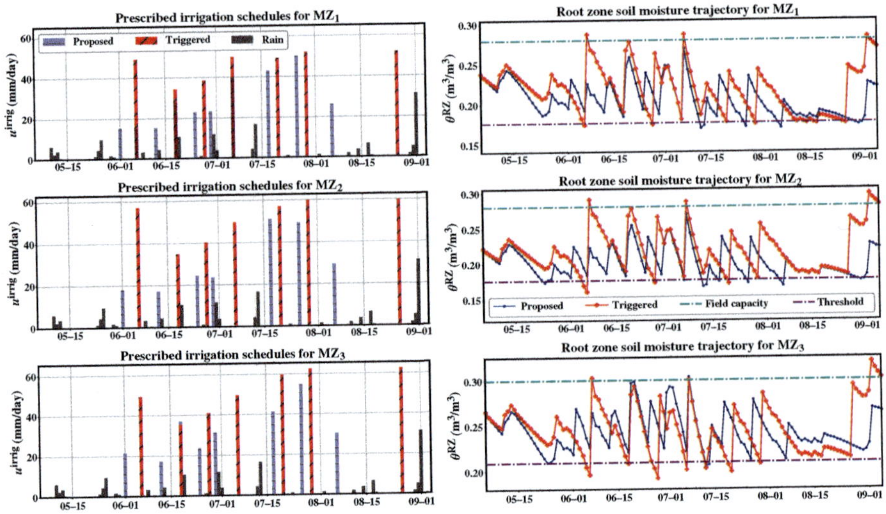


**Figure 7.10** Irrigation schedules and resulting root zone soil moisture trajectories for the 2022 wet season under a 65% MAD scenario. *Source*: [105] with permission of Elsevier/CC BY 4.0.

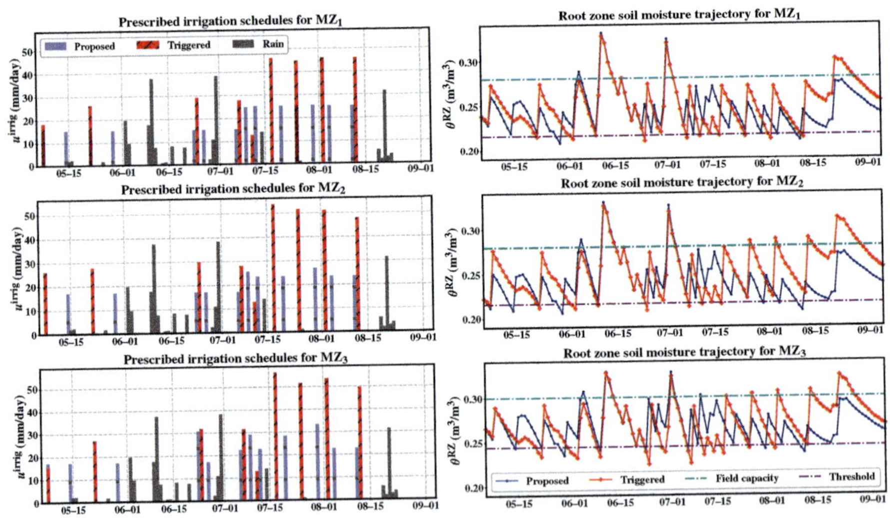


**Figure 7.11** Irrigation schedules and resulting root zone soil moisture trajectories for the 2015 dry season under a 40% MAD scenario. *Source*: [105] with permission of Elsevier/CC BY 4.0.

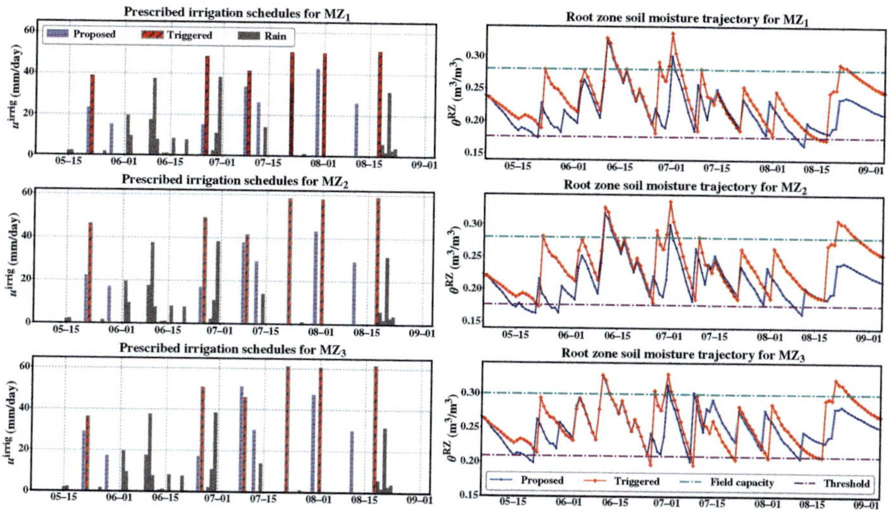


**Figure 7.12** Irrigation schedules and resulting root zone soil moisture trajectories for the 2015 dry season under a 65% MAD scenario. *Source*: [105] with permission of Elsevier/CC BY 4.0.

**Table 7.2** Triggered versus proposed irrigation scheduling for the 2022 season under a conservative MAD of 40%.

| | Proposed | Triggered |
|---|---|---|
| Irrigation for $MZ_1$ (m) | 0.28 | 0.29 |
| Irrigation for $MZ_2$ (m) | 0.29 | 0.33 |
| Irrigation for $MZ_3$ (m) | 0.32 | 0.33 |
| Total irrigation (m) | 0.89 (−6%) | 0.95 |
| IWUE ($kg/m^3$) | 2.83 (+12%) | 2.53 |

**Table 7.3** Triggered versus proposed irrigation scheduling for the 2022 season under a less conservative MAD of 65%.

| | Proposed | Triggered |
|---|---|---|
| Irrigation for $MZ_1$ (m) | 0.22 | 0.28 |
| Irrigation for $MZ_2$ (m) | 0.23 | 0.31 |
| Irrigation for $MZ_3$ (m) | 0.26 | 0.32 |
| Total irrigation (m) | 0.71 (−22%) | 0.91 |
| IWUE ($kg/m^3$) | 3.60 (+35%) | 2.66 |

**Table 7.4** Triggered versus proposed irrigation scheduling for the 2015 season under a conservative MAD of 40%.

| | Proposed | Triggered |
|---|---|---|
| Irrigation for $MZ_1$ (m) | 0.33 | 0.36 |
| Irrigation for $MZ_2$ (m) | 0.34 | 0.39 |
| Irrigation for $MZ_3$ (m) | 0.38 | 0.38 |
| Total irrigation (m) | 1.05 (−7%) | 1.13 |
| IWUE (kg/m$^3$) | 2.41 (+10%) | 2.20 |

**Table 7.5** Triggered versus proposed irrigation scheduling for the 2015 season under a less conservative MAD of 65%.

| | Proposed | Triggered |
|---|---|---|
| Irrigation for $MZ_1$ (m) | 0.28 | 0.32 |
| Irrigation for $MZ_2$ (m) | 0.30 | 0.36 |
| Irrigation for $MZ_3$ (m) | 0.31 | 0.36 |
| Total irrigation (m) | 0.89 (−14%) | 1.04 |
| IWUE (kg/m$^3$) | 2.96 (+20%) | 2.46 |

## 7.7 Conclusion

This chapter developed a integrated irrigation scheduling framework that combines machine learning, MPC and multi-agent systems. The framework is designed to address the challenges of efficient irrigation scheduling in large-scale agricultural fields that exhibit spatial heterogeneity. The process began with a three-stage approach to MZ delineation. In this delineation, $k$-means clustering was applied to topographical and estimated hydraulic data to partition the field into zones with similar agro-hydrological characteristics. These zones were then refined to align with the spatial resolution of the irrigation system, enabling efficient water application. To model soil moisture dynamics within the delineated zones, LSTM networks were trained using data generated from calibrated 1D Richards' equations. These surrogate models accounted for variations in rooting depth and provided reliable one-day-ahead predictions of root zone moisture. The trained LSTM networks were then embedded in an MPC framework to compute optimal daily irrigation strategies that balanced moisture regulation with efficient resource (water, labor, and energy) use. Recognizing the computational burden of solving mixed-integer MPC problems, particularly in multi-MZ settings, the framework introduced a decentralized learning approach. Each MZ was assigned a hybrid RL agent trained using the PPO algorithm. To harmonize binary irrigation decisions across zones, a limiting MZ strategy was used to guide field-wide coordination. This enabled the decomposition of the scheduling problem into simpler and independent subproblems that could be solved efficiently and in parallel. Evaluation of the framework across different growing seasons demonstrated its ability to reduce water use while improving IWUE compared to a triggered approach.

# 8

# Optimizing Irrigation with Semi-centralized Multi-agent RL

## 8.1 Introduction

Chapter 7 demonstrated the application of decentralized multi-agent reinforcement learning (MARL) to improve the computational performance of mixed-integer model predictive control (MPC) for irrigation scheduling in heterogeneous fields. In that framework, MARL agents were trained offline to generate binary irrigation decisions independently of the MPC. This reduced the scheduling problem from a mixed-integer nonlinear program to a nonlinear program. Additionally, the decentralized architecture supported parallel computation, enabling simultaneous evaluation of irrigation amounts across different management zones (MZs).

Despite these advantages, the decentralized MARL (DMARL) method presented a key challenge related to coordination. In the original optimal control formulation, the binary irrigation decision is applied uniformly across all zones, making it a global decision. When local agents determine this decision independently and rely on heuristic methods for coordination, the resulting uniform action may not reflect a globally optimal choice. This gap can limit the system's ability to deliver high scheduling accuracy and water-use efficiency.

To improve coordination while maintaining scalability, this chapter adopts a semi-centralized MARL framework. The framework divides the irrigation scheduling process into two decision-making levels, combining the strengths of centralized control with decentralized decision-making:

1. **Centralized Coordinator Agent:** This agent oversees high-level binary irrigation timing decisions. It operates using field-wide information (e.g., overall soil moisture patterns, weather predictions, and crop information) to determine whether irrigation should occur on a given day. Centralizing this decision ensures consistency across the field and helps avoid the suboptimal outcomes that may result from aggregating decisions made independently by local agents.
2. **Decentralized Local Agents:** These agents are responsible for determining the irrigation amounts applied within their assigned MZs. Drawing on local data, such as zone-specific moisture levels, crop types, and weather conditions, they operate independently to respond to spatial differences in field conditions. This decentralized layer preserves the scalability and adaptability of the MARL approach discussed in Chapter 7.

*Precision Irrigation for Agriculture: Integrating Machine Learning and Optimal Control Strategies*, First Edition. Bernard Twum Agyeman and Jinfeng Liu.

A common occurrence in MARL is the presence of non-stationarity. When several agents update their policies concurrently, the environment experienced by each agent changes over time due to the evolving actions of others. This results in non-stationary transition dynamics and reward signals, which can hinder learning stability.

In the DMARL framework discussed in Chapter 7, the issue of non-stationarity was avoided by training agents independently. In contrast, the semi-centralized MARL framework introduces interaction between a centralized coordinator and decentralized local agents, reintroducing the challenge of non-stationarity. As the coordinator influences the decision-making context for all local agents, the learning environment becomes dynamic and more difficult to model. A state augmentation strategy is leveraged to mitigate the issue of non-stationarity. This approach incorporates relevant global information into the observation space of local agents, allowing them to adapt to changes introduced by the centralized coordinator while preserving the benefits of decentralized learning.

The effectiveness of the semi-centralized MARL framework is demonstrated through its application to a large-scale agricultural field, where its performance is benchmarked against the DMARL approach presented in Chapter 7. Results, which were originally reported in [109], highlight improvements in coordination, scheduling accuracy, and overall irrigation water-use efficiency.

## 8.2 Semi-centralized MARL (SCMARL) Framework

To manage the complexity associated with irrigation scheduling in spatially heterogeneous agricultural fields, an SCMARL framework is introduced. The framework is designed for daily scheduling in large-scale fields divided into multiple MZs. For an in-depth explanation of MARL principles and training methods, readers are referred to Section 3.1.4 of Chapter 3.

The SCMARL framework adopts a hierarchical structure, as illustrated in Figure 8.1, comprising a centralized coordinator agent and a set of decentralized local agents. Each agent is responsible for a specific role in the scheduling process, enabling coordinated decision-making at the field scale while accommodating localized variability within individual MZs.

At the upper level, the coordinator agent is responsible for issuing a daily binary irrigation decision—either to irrigate ("yes/1") or not ("no/0"). This decision is based on field-wide data, including soil moisture conditions of the MZs, crop-related variables and weather forecasts. The binary output serves as a control signal that enables or disables the irrigation amounts determined by local agents. A "yes/1" decision activates irrigation across all zones, while a "no/0" decision suppresses irrigation throughout the field.

At the lower level, local agents are responsible for determining the irrigation amounts required within their respective MZs. These agents rely on zone-specific inputs, such as local soil moisture levels, crop-related variables and weather conditions, in combination with the binary irrigation decision issued by the coordinator. The final irrigation amount implemented in each zone is obtained by multiplying the local agent's recommendation by the coordinator's binary decision in order to ensure alignment with the field-level irrigation directive.

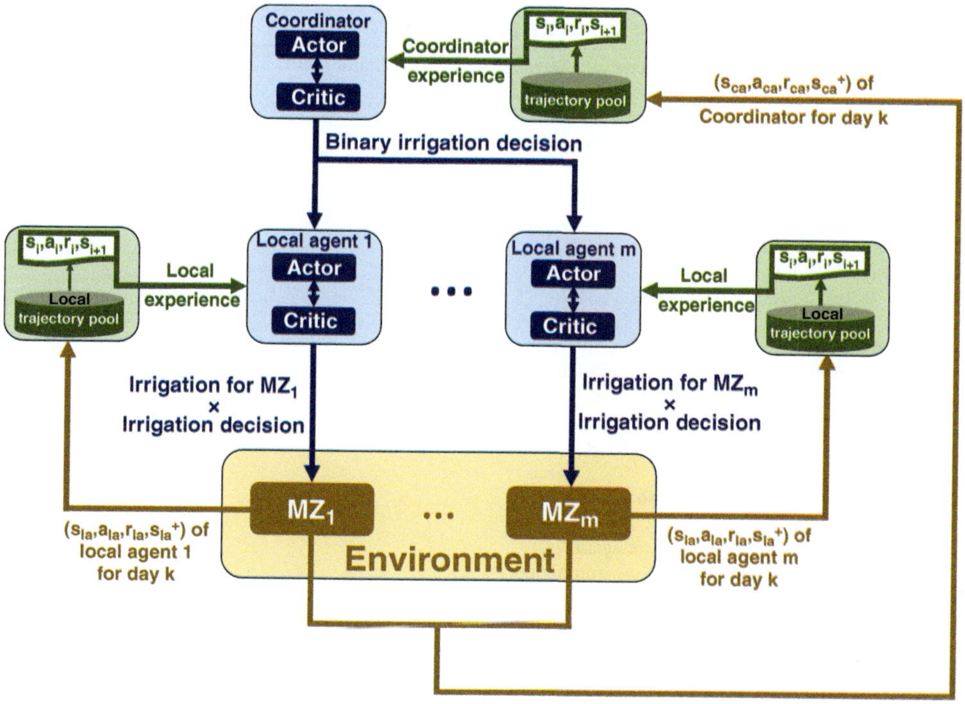


**Figure 8.1** Schematic representation of the semi-centralized MARL framework used for irrigation scheduling. *Source:* [109] with permission of Elsevier.

Although various RL techniques can be employed within this framework, this chapter adopts the proximal policy optimization (PPO) algorithm, a widely used actor-critic method known for its stability and robustness. A comprehensive explanation of PPO is provided in Section 3.1.3.5.2 of Chapter 3. In the training phase, each agent collects data by interacting with its own environment, storing the resulting experience in a trajectory buffer. This information is later used to refine both the actor and critic components of the agent. The learning process involves optimizing a penalized objective that incorporates *advantage* estimates to guide updates. To maintain training stability and avoid drastic shifts in the policy, PPO introduces a clipping mechanism that constrains how much the policy can be changed in a single update. Experience samples are periodically drawn from each agent's buffer to update its neural networks. The full scheduling procedure based on the SCMARL framework is outlined in Algorithm 3.

### 8.2.1 Addressing Non-stationarity in SCMARL Frameworks

As highlighted earlier, the framework builds on the observation that interactions between MZs in large-scale fields are relatively weak, and therefore models each MZ as an independent unit. This design choice simplifies the system by reducing inter-agent dependencies at the lower level, allowing each local agent to operate with minimal influence on the rewards and state transitions of others.

**Algorithm 3** Pseudocode of the SCMARL Algorithm

Initialize weights of critic $C_{\text{ca}}$ and actor $\mathcal{A}_{\text{ca}}$ networks of the coordinator agent.
**for** local agent $i \leftarrow 1$ to $M$ **do**
Initialize weights of critic $C_{\text{la}_i}$ networks of $i$.
Initialize weights of actor $\mathcal{A}_{\text{la}_i}$ networks of $i$.
**end for**
Set the total number of episodes $K$.
Set the horizon length for each episode $T$.
**for** episode $k \leftarrow 1$ to $K$ **do**
**for** local agent $i \leftarrow 1$ to $M$ **do**
Obtain initial soil moisture ($y_{i,\text{init}}$) of $i$.
Set $y_i \leftarrow y_{i,\text{init}}$.
**end for**
Obtain initial weather and crop coefficient data $\text{ET}_{0,\text{init}}, R_{n,\text{init}}, K_{c,\text{int}}$.
Set $\text{ET}_0 \leftarrow \text{ET}_{0,\text{init}}, R_n \leftarrow R_{n,\text{int}}, K_c \leftarrow K_{c,\text{int}}$.
**for** timestep $t \leftarrow 1$ to $T$ **do**
Obtain $s_{\text{ca}} = [y_1, \ldots, y_M, \text{ET}_0, K_c, R_n]$.
Select $a_{\text{ca}} = c = \mathcal{A}_{\text{ca}}(s_{\text{ca}})$.
**for** local agent $i \leftarrow 1$ to $M$ **do**
Obtain $s_{\text{la}_i} = [y_i, \text{ET}_0, K_c, R_n, c]$.
Obtain $a_{\text{la}_i} = \mathcal{A}_{\text{la}_i}(s_{\text{la}_i})$ for agent $i$.
Execute $u^{\text{irr}}_{\text{la}_i} = a_{\text{la}_i} \times c$ in environment of $i$.
Obtain $y_i^+$ from Equations (7.4) and (7.5).
Obtain $r_{\text{la}_i}$ from Equation (8.1).
**end for**
Store $[s_{\text{la}_1}, \ldots, s_{\text{la}_M}], [a_{\text{la}_1}, \ldots, a_{\text{la}_M}], [r_{\text{la}_1}, \ldots, r_{\text{la}_M}]$, and $[y^+_{\text{la}_1}, \ldots, y^+_{\text{la}_M}]$.
Obtain weather and crop coefficient predictions as $\text{ET}_0^+, R_n^+, K_c^+$.
Obtain $s^+_{\text{ca}} = [y_1^+, \ldots, y_M^+, \text{ET}_0^+, K_c^+, R_n^+]$.
Obtain $r_{\text{ca}}$ from Equation (8.4).
Store $(s_{\text{ca}}, a_{\text{ca}}, r_{\text{ca}}, s^+_{\text{ca}})$ for coordinator.
Update $C_{\text{ca}}$ and $\mathcal{A}_{\text{ca}}$.
Obtain $c^+ = \mathcal{A}_{\text{ca}}(s^+_{\text{ca}})$.
**for** local agent $i \leftarrow 1$ to $M$ **do**
Obtain $s^+_{\text{la}_i} = [y_i^+, \text{ET}_0^+, K_c^+, R_n^+, c^+]$.
Store $(s_{\text{la}_i}, a_{\text{la}_i}, r_{\text{la}_i}, s^+_{\text{la}_i})$ for $i$.
Update $C_{\text{la}_i}$ and $\mathcal{A}_{\text{la}_i}$ for $i$.
**end for**
**for** local agent $i \leftarrow 1$ to $M$ **do**
Set $y_i \leftarrow y_i^+$.
**end for**
Set $\text{ET}_0 \leftarrow \text{ET}_0^+, R_n \leftarrow R_n^+, K_c \leftarrow K_c^+$.
**end for**
**end for**

However, to ensure alignment with the overall field-wide irrigation strategy, the coordinator's action is used to activate or deactivate irrigation across all MZs. While this mechanism supports consistency with the scheduling objective, it also introduces a key source of non-stationarity into the learning environments of the local agents. Each local agent must now learn in an environment where outcomes depend not only on its own actions but also on the coordinator's decision. This added variability, which is beyond the agent's control, increases the complexity of the learning process and makes it more difficult for local agents to develop stable policies.

To promote more stable learning at the level of the local agents, the framework adopts two main strategies:

1. **Communication of Coordinator's Action:** Agents at the local level receive the coordinator's action before selecting their actions. The local agents are able to access this information since the coordinator acts before they choose their actions.
2. **State Augmentation:** Each local agent's input includes both the coordinator's action and relevant zone-level data. Consequently, the local agents are informed of the field-wide irrigation directive at the time of decision-making. By using this feed-forward structure, agents are able to align their actions with the coordinator's action and adjust their policies accordingly.

Moreover, integrating the coordinator's action into the input of each local agent and enforcing this decision in a switch-like manner during training facilitates alignment between local recommendations and the field-wide irrigation decision. Through reward-based learning, local agents gradually adapt to output nonzero irrigation amounts when the coordinator issues a "yes" action, and zero amounts when the action is "no." This learned behavior reduces the need for manual enforcement of the binary decision during execution, as local agents naturally adjust their actions to follow the coordinator's action.

Although the coordinator agent's environment is shaped by the irrigation amounts proposed by the local agents, it ultimately determines whether irrigation is activated across the field. This centralized authority ensures consistent enforcement of the action of the coordinator. Consequently, the coordinator experiences less variability in the outcomes of its action, leading to a more stable learning environment compared to that of the local agents.

## 8.2.2 Design of Local Agents

As previously noted, each local agent is responsible for deciding how much irrigation should be applied each day within its respective MZ. The following sections describe the key components of the Markov decision process (MDP) that inform the local agent's decision-making process.

### 8.2.2.1 Environmental Modeling for Local Agents

Each agent at the local level models soil moisture behavior using a one-dimensional (1D) version of the Richards' equation. To ensure that this model reliably captures zone-specific behavior, it must be calibrated by assigning appropriate values to the parameter $\phi$ in Equations (7.4) and (7.5) in Chapter 7. This calibration assumes that soil properties are uniform within each MZ. Parameter estimation approaches, such as the one discussed in Chapter 4, can be used to identify parameter values that accurately represent the physical characteristics of the soil in each zone.

### 8.2.2.2 State Representation for Local Agents

The soil moisture across the spatial domain, $y$, serves as the primary input to an agent at the local level. This primary input is augmented with several variables that are essential for irrigation scheduling, including daily values of reference evapotranspiration ($\mathrm{ET}_0$), crop coefficient ($K_c$), and precipitation ($R_n$). The action ($c$), issued by the coordinator agent for the current day, is also incorporated. The resulting state representation is given by $s_{\mathrm{la}} := [y, \mathrm{ET}_0, K_c, R_n, c]$.

Given $s_{\text{la}}$, the actor network $\mathcal{A}_{\text{la}}$ of the local agent computes the irrigation amount $a_{\text{la}}$ corresponding to the agent's MZ. This value is multiplied by $c$ to obtain the final action, defined as $u_{\text{la}}^{\text{irrig}} := a_{\text{la}} \times c$. Applying this action to the environment updates the soil moisture state to $y^+$, in accordance with Equations (7.4) and (7.5) in Chapter 7.

To construct the next state $s_{\text{la}}^+$, the agent incorporates predicted values of $\text{ET}_0^+$, $K_c^+$, and $R_n^+$ for the following day. The coordinator's updated actor $\mathcal{A}_{\text{ca}}$ is then used to determine the next binary irrigation action $c^+$, based on a global input aggregated from all MZs. This input includes the predicted soil moisture levels $y^+$ and the forecasted environmental and crop variables (i.e., $\text{ET}_0^+$, $K_c^+$, and $R_n^+$), forming the coordinator's state $s_{\text{ca}}^+ := [y_1^+, y_2^+, \ldots, y_M^+, \text{ET}_0^+, K_c^+, R_n^+]$. The decision is computed as $c^+ := \mathcal{A}_{\text{ca}}(s_{\text{ca}}^+)$, after which the local agent updates its state to $s_{\text{la}}^+ := [y^+, \text{ET}_0^+, K_c^+, R_n^+, c^+]$.

#### 8.2.2.3 Reward Function for Local Agents

The agent's decision-making process is guided by two objectives. The primary objective is to maintain the root zone soil moisture content, $\theta^{\text{RZ}}$, within a target range defined by lower and upper bounds, $\underline{\nu}$ and $\overline{\nu}$, respectively. These bounds correspond to the target zone introduced in Chapter 7 and are computed using Equations (7.8) and (7.9). The root zone moisture content $\theta^{\text{RZ}}$ is estimated using the weighted method outlined in Section 7.3.2 of Chapter 7, which integrates moisture values from different soil layers in the MZ.

A secondary objective is to reduce the amount of water applied, thereby encouraging more efficient use of water resources. These two objectives are jointly represented in the local agent's reward function, $r_{\text{la}}$, which combines a target-zone tracking term, $r_{\text{la}}^z$, and an irrigation minimization term, $r_{\text{la}}^u$:

$$r_{\text{la}} := \alpha_{\text{la}} r_{\text{la}}^z + \beta_{\text{la}} r_{\text{la}}^u \tag{8.1}$$

The parameters $\alpha_{\text{la}}$ and $\beta_{\text{la}}$ specify the relative importance of maintaining soil moisture within the target zone and minimizing irrigation usage, respectively. The term $r_{\text{la}}^z$, which quantifies how well the agent keeps $\theta^{\text{RZ}}$ within the bounds $[\underline{\nu}, \overline{\nu}]$, is defined as follows:

$$r_{\text{la}}^z := \begin{cases} -\underline{Q} \times |\theta^{\text{RZ}} - \underline{\nu}| & \text{if } \theta^{\text{RZ}} < \underline{\nu} \\ -\overline{Q} \times |\theta^{\text{RZ}} - \overline{\nu}| & \text{if } \theta^{\text{RZ}} > \overline{\nu} \\ 0 & \text{if } \underline{\nu} \le \theta^{\text{RZ}} \le \overline{\nu} \end{cases} \tag{8.2}$$

The terms $\underline{Q} > 0$ and $\overline{Q} > 0$ serve to penalize the deviations below and above the moisture target range, respectively. Notably, $\theta^{\text{RZ}}$, is computed using $y^+$, which, as previously noted, reflects the outcome of applying the local agent's action after it has been modulated by the coordinator's binary decision.

As defined in Equation (8.2), the local agent is penalized proportionally to its deviation from the target range when the root zone soil moisture content $\theta_k^{\text{RZ}}$ lies outside the bounds. A penalty of $-\underline{Q} \times |\theta_k^{\text{RZ}} - \underline{\nu}|$ is applied when $\theta_k^{\text{RZ}} < \underline{\nu}$, and $-\overline{Q} \times |\theta_k^{\text{RZ}} - \overline{\nu}|$ when $\theta_k^{\text{RZ}} > \overline{\nu}$. No penalty is incurred when $\theta_k^{\text{RZ}}$ remains within $[\underline{\nu}, \overline{\nu}]$, yielding a reward of zero.

The component of the reward that penalizes irrigation usage, denoted $r_{\text{la}}^u$, is expressed as:

$$r_{\text{la}}^u := -R_u a_{\text{la}} \tag{8.3}$$

where $R_u > 0$ denotes the cost assigned to each unit of water applied.

### 8.2.3 Design of the Coordinator Agent

As previously discussed, the coordinator utilizes spatially aggregated soil moisture, crop, and weather variables to determine whether to irrigate ($c$) on a particular day. This section elaborates on the essential elements of the MDP that guide the design of the coordinator.

#### 8.2.3.1 Environmental Dynamics of the Coordinator

The coordinator's training is based on simulations generated using $M$ distinct 1D Richards' equations, each calibrated to reflect the soil hydraulic properties of a specific MZ. These models (Equations 6.10 and 6.11 in Chapter 6) incorporate localized soil characteristics through their respective calibration parameters.

This modeling approach assumes weak coupling between MZs, allowing the coordinator to treat the field-level dynamics as a concatenation of the individual zone-level models. Since these models are already defined during the training of the local agents, there is no need to respecify them for the coordinator. Instead, the coordinator reuses the calibrated equations for the local agents to simulate soil moisture evolution across the field.

#### 8.2.3.2 State Representation for the Coordinator

To represent field-wide conditions, the soil moisture vectors from all $M$ MZs are combined into a single input, $Y := [y_1, y_2, \ldots, y_M]$. This vector is concatenated with the daily values of weather and crop variables to define the state observed by the coordinator: $s_{\text{ca}} := [Y, \text{ET}_0, K_c, R_n]$. Using $s_{\text{ca}}$, the actor $\mathcal{A}_{\text{ca}}$ of the coordinator generates $c$. This results in a coordinated control input across the field, represented as $A := [c \times a_{\text{la}_1}, c \times a_{\text{la}_2}, \ldots, c \times a_{\text{la}_M}]$. When $A$ is implemented, the system evolves to a new spatial soil moisture configuration $Y^+$, governed by Equations (6.10) and (6.11) in Chapter 6. The updated version of $s_{\text{ca}}$ ($s_{\text{ca}}^+$), is constructed by combining the successor soil moisture vector $Y^+$ with the next day's predicted crop and weather variables (i.e., $\text{ET}_0^+$, $K_c^+$, and $R_n^+$) and this yields $s_{\text{ca}}^+ := [Y^+, \text{ET}_0^+, K_c^+, R_n^+]$.

#### 8.2.3.3 Reward Function for the Coordinator

The coordinator is tasked with finding a daily binary irrigation decision that maintains the root zone soil moisture content in each MZ within a specified moisture range. In addition to this soil moisture regulation objective, the agent aims to minimize the fixed costs incurred when an irrigation event is performed. Accordingly, this agent's reward, $r_{\text{ca}}$, is composed of two components: the moisture regulation reward, $r_{\text{ca}}^z$ and the cost minimization reward, $r_{\text{ca}}^c$:

$$r_{\text{ca}} := \alpha_{\text{ca}} r_{\text{ca}}^z + \beta_{\text{ca}} r_{\text{ca}}^c \tag{8.4}$$

where $\alpha_{\text{ca}}$ and $\beta_{\text{ca}}$ are weighting parameters that reflect the relative priorities assigned to moisture regulation and cost minimization.

The term $r_{\text{ca}}^z$ captures how well root zone soil moisture levels across the field align with their respective ranges. It is computed as the aggregate of individual zone-level rewards:

$$r_{\text{ca}}^z := \sum_{i=1}^{M} r_{\text{la},i}^z \tag{8.5}$$

In Equation (8.5), $r^{z}_{\text{la},i}$ denotes the moisture tracking reward for MZ $i$, as defined by Equation (8.2). Furthermore, the term $r^{c}_{\text{ca}}$ captures the cost incurred when irrigation is initiated:

$$r^{c}_{\text{ca}} := -R_c c \tag{8.6}$$

The positive parameter $R_c$ in Equation (8.6) defines the cost associated with executing the irrigation decision and is applied whenever $c = 1$.

## 8.3 Field-scale Implementation of the SCMARL Framework

Quadrant 3, which was first introduced in Chapter 7 as part of a larger study area, was selected for the implementation of the SCMARL. The quadrant in question was partitioned into three MZs using a delineation method that was guided by spatial variability in soil hydraulic characteristics and elevation. Moreover, the hydraulic parameters employed for the delineation were estimated through the data assimilation technique described in Chapter 4.

Each zone's soil moisture behavior was represented using a calibrated 1D Richards' equation, with the calibration based on the hydraulic properties at the centroid of the zone. During the study period, wheat was cultivated uniformly throughout the quadrant. The SCMARL architecture consisted of four agents: a coordinator responsible for quadrant-wide irrigation control and three local agents, each dedicated to an individual MZ.

### 8.3.1 Simulation Environment Configuration

Each MZ was simulated using a vertical soil profile of 0.50 m, partitioned into 21 evenly spaced layers. The simulation horizon spanned 24 hours, with a fixed time step of 30 minutes to ensure accurate integration. Initial moisture conditions were derived from the final soil moisture distributions obtained during the estimation of soil moisture and hydraulic parameters, which was completed prior to the delineation of the MZs.

Environmental conditions during training were derived from historical data. Evapotranspiration values, on a daily basis, were sampled uniformly within the range of 1.04–9.0 mm/day. Daily values of rainfall and temperature were drawn from a dataset comprising 2,448 days, corresponding to the growing seasons from 2005 to 2020 (May to September). Crop coefficients for wheat were computed using a temperature-based empirical relation (see Equation (A.3), Appendix (A.4)). A rooting depth of 0.50 m was applied consistently, in line with typical wheat irrigation practices.

To quantify uncertainty in both system behavior and environmental observations, Gaussian noise was introduced. Specifically, the noise terms $\omega$ and $\nu$ in Equations (7.4) and (7.5) were sampled from zero-mean normal distributions with standard deviations of 0.0002 and 0.0005, respectively. Additional uncertainty was added to the environmental and crop inputs available to agents during training. While agents used these perturbed values for decision-making, soil moisture dynamics were updated using the original inputs.

### 8.3.2 Agent Setup and Training Process

The reward function parameters were assigned distinctly for the local and coordinator agents. For the local agents, $\alpha_{la} = 1.0$ and $\beta_{la} = 1.0$ were used, while the coordinator agent employed $\alpha_{ca} = 0.1$ and $\beta_{ca} = 1.0$. The values of $\bar{Q}$ and $\underline{Q}$ were fixed at 1,200,000 and 1,000,000, respectively. Other penalty coefficients were set to $R_c = 1,000$ and $R_u = 9,000$. A MAD value of 50% was employed to calculate the bounds of the soil moisture ranges. With respect to $MZ_1$ and $MZ_2$, the bounds were $\bar{\nu} = 0.280$ and $\underline{\nu} = 0.200$, respectively, while in $MZ_3$, they were $\bar{\nu} = 0.300$ and $\underline{\nu} = 0.230$. The parameters $\theta_{fc}$ and $\theta_{wp}$ used in these calculations were obtained from [106].

A learning rate of $1 \times 10^{-5}$ was employed to train the actor and critic networks. Input features were scaled to the range $[-2.0, 2.0]$ using min-max normalization. Each episode spanned a 30-day horizon. This horizon was selected based on simulation insights that identified it as an effective compromise between short-term adaptability and long-term scheduling performance. Training employed a minibatch size of 64, ran for 20 epochs, and used a discount factor of 0.99, a generalized advantage estimation coefficient of 0.97, a clipping threshold of 0.25, and an entropy regularization term set to 0.01.

The coordinator operated in a discrete action space and used a softmax distribution for action sampling. In contrast, local agents generated continuous irrigation decisions through Gaussian distributions. All agents used policy networks consisting of two fully connected hidden layers with 64 units each and tanh activation. Following the SCMARL configuration, the input dimension was 25 for each local agent and 66 for the coordinator agent. The agents were implemented and trained using the Tensorforce library [107], over a total of 9,800 episodes distributed across 10 training runs. Importantly, all parameters described in this section were treated as tunable and were iteratively refined through experimentation until acceptable performance was achieved.

### 8.3.3 Learning Outcomes of Agents

The results in Figure 8.2 summarize how agent performance changes over the course of training. Each curve represents the average score from 10 separate runs, covering 9,800 episodes per agent. For roughly the first 6,000 episodes, the agents show consistent improvement, as seen in the upward movement of the score curves. After this phase, the curves flatten, indicating that further training does not lead to substantial performance gains. This pattern suggests that the agents reach a point where their actions no longer change significantly and their performance stabilizes. These results show that agents trained under the SCMARL framework are able to reach stable behavior through repeated interaction with the environment.

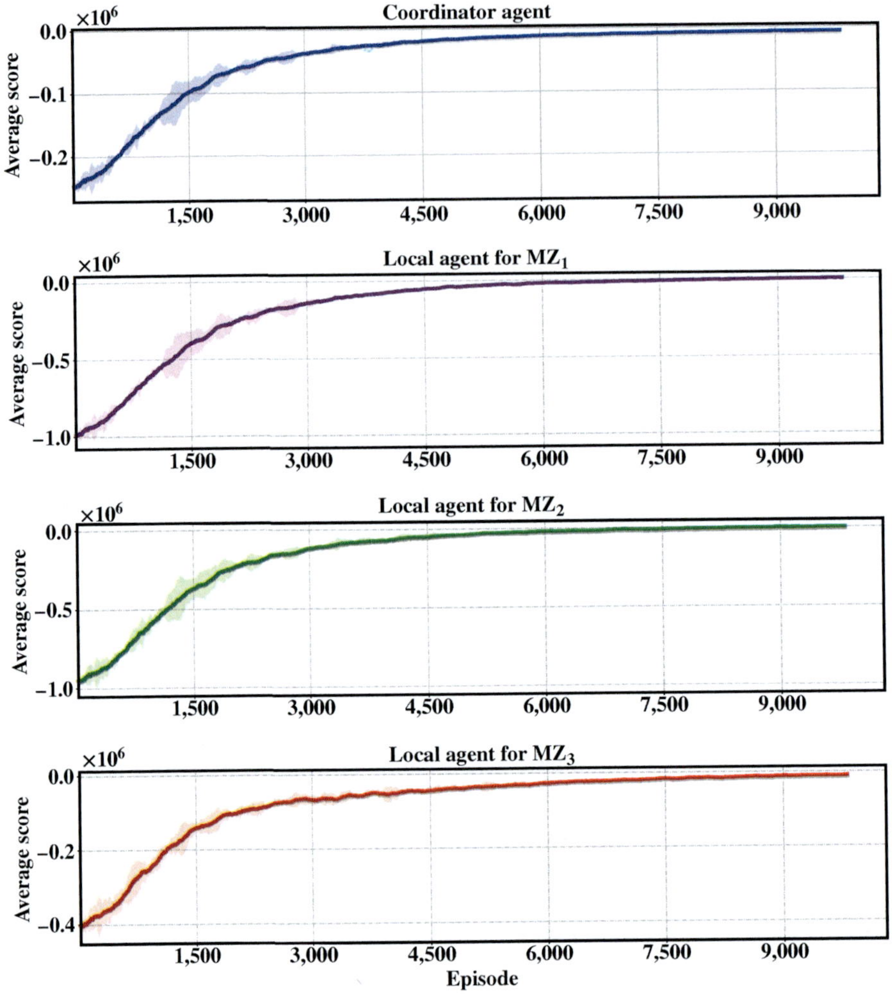


**Figure 8.2** Evolution of the average score throughout 9,800 episodes, averaged over 10 independent training runs. *Source:* Adapted from [109].

## 8.4 Evaluation of SCMARL

Three key experiments were carried out to assess the performance of the SCMARL framework. The first focused on evaluating how the inclusion of the coordinator agent's decisions influences the learning behavior and overall performance of the agents at the local level. The second explored the extent to which local agents can learn policies that align with the global decisions provided by the coordinator. The third experiment focused on comparing the performance of the SCMARL framework with that of the DMARL approach introduced in Chapter 7.

### 8.4.1 Evaluation of the State Augmentation Strategy

To evaluate the role of communication and state augmentation in the SCMARL framework, a comparison was made between the proposed approach and a modified version that excludes state augmentation. In the modified setup, a coordinator still generated binary irrigation actions based on global inputs, but these actions were neither shared with the local agents nor reflected in their observation space. As a result, the coordinator operated independently, serving only to permit or block the irrigation levels selected by the local agents.

The comparison focused on two aspects. The first comparison involved analyzing the learning performance of each framework by tracking mean score trends over 9,800 training episodes and calculating the mean training time, based on results from 10 independent runs. The second comparison tested a practical application of both methods by using them to generate irrigation schedules for wheat during the 2022 growing season. Relevant weather conditions for this period are illustrated in Figure A.3 of Appendix (A.3). Performance was further assessed using two key metrics: the total amount of irrigation applied and the irrigation water use efficiency (IWUE), as defined in Equation (7.14). The subsequent sections provide a detailed overview of the experimental conditions and implementation details.

#### 8.4.1.1 Training Conditions for the Non-augmented SCMARL Variant

The same training settings outlined in Section 8.3.2 were applied when training agents in both the original SCMARL framework and the version without state augmentation. To maintain consistency across experiments, the same weight initialization technique was also used in both cases. This ensured that any differences in performance could be attributed to the presence or absence of state augmentation rather than variations in training configuration.

#### 8.4.1.2 Experimental Settings for the Full-season Evaluation

Although both variants of the SCMARL framework were trained under the assumption of full observability, the season-long evaluation was designed to reflect a partially observable setting. In this setting, it was considered that detailed soil moisture profiles within each MZ were not directly accessible. To fill this gap, simulated remote sensing data provided daily readings of the mean moisture content in the upper 25 cm of the soil profile for each zone. These sparse (with respect to space) measurements were then used to reconstruct the full moisture distribution using an extended Kalman filter (EKF). Additional details regarding the design of EKF can be found in Appendix D.

The field-scale evaluation covered the period from May 15, 2022 to September 4, 2022. At the start of this evaluation, initial soil moisture estimates were assigned to each of the three MZs. These estimates, together with daily crop and weather data, provided the basis for calculating schedules on the first day. The resulting prescriptions guided water distribution within a simulated field system consisting of three independently calibrated 1D Richards' equation models, one for each MZ. Each soil profile extended to a depth of 0.5 m and was divided into 21 uniform layers. To capture uncertainty in the simulated field environment, Gaussian process noise with a mean of 0 and a standard deviation of 0.0007 was introduced during the simulation.

On day two of the evaluation, each MZ received a sensor reading that reflected the average water content in the top 25 cm of its soil column. To model sensor uncertainty, Gaussian noise with a mean of 0 and a standard deviation of 0.0008 was added to the measurements. For each MZ, the noisy and sparse reading was used to refine the moisture estimates using the EKF. Furthermore, random noise was introduced into the crop and weather inputs to account for uncertainty in these uncontrolled inputs. The refined moisture estimates across MZs, together with the perturbed uncontrolled inputs, were then used to determine the quadrant-wide "yes/no" decision and the corresponding irrigation quantities for all the MZs. This process was repeated daily for the duration of the simulation.

### 8.4.2 Local Agent Policy Alignment with Coordinator

The purpose of this analysis was to evaluate how closely the local agents' actions followed the coordinator's "yes/no" decisions under both SCMARL configurations. Local agent behavior was considered aligned with the coordinator if the following criteria were met:

1. A zero irrigation amount was applied when the coordinator issued a "no" decision.
2. A nonzero irrigation amount was applied when the coordinator issued a "yes" decision.

The evaluation consisted of 2,000 trials per configuration, during which the rate of alignment between local agent actions and coordinator decisions was recorded.

### 8.4.3 SCMARL Versus DMARL: A Comparative Study

This comparison involved applying both the SCMARL framework and the DMARL approach with MPC, previously introduced in Chapter 7, to generate prescriptions for wheat cultivation in Quadrant 3. The evaluation focused on two metrics: total prescribed irrigation and IWUE. The sections that follow describe the parameters and experimental setup used for this study.

#### 8.4.3.1 Training of Hybrid PPO Agents in the DMARL

To ensure comparability between the two scheduling techniques, the DMARL agents were trained using the same reward weighting scheme as the SCMARL agents. The training process also adopted the hyperparameter configuration provided in Section 8.3.2. Additionally, the neural network initialization procedure used for SCMARL was applied to the DMARL agents to maintain consistency in initial training conditions.

#### 8.4.3.2 Evaluation Setting of the SCMARL

For a meaningful performance benchmark against the DMARL approach described in Chapter 7, the daily evaluation procedure for SCMARL agents was modified. This modified evaluation was conducted over a fixed scheduling horizon, consistent with the framework used for DMARL. An overview of this procedure is provided in Figure 8.3. The modified evaluation began with the initial soil moisture profiles in each MZ, alongside predicted uncontrolled inputs (i.e., weather and crop data forecasts) over the entire horizon. Based on this input data, the SCMARL agents interacted with their respective environments to generate a sequence of "yes/no" actions and the associated irrigation amounts for each MZ across the horizon.

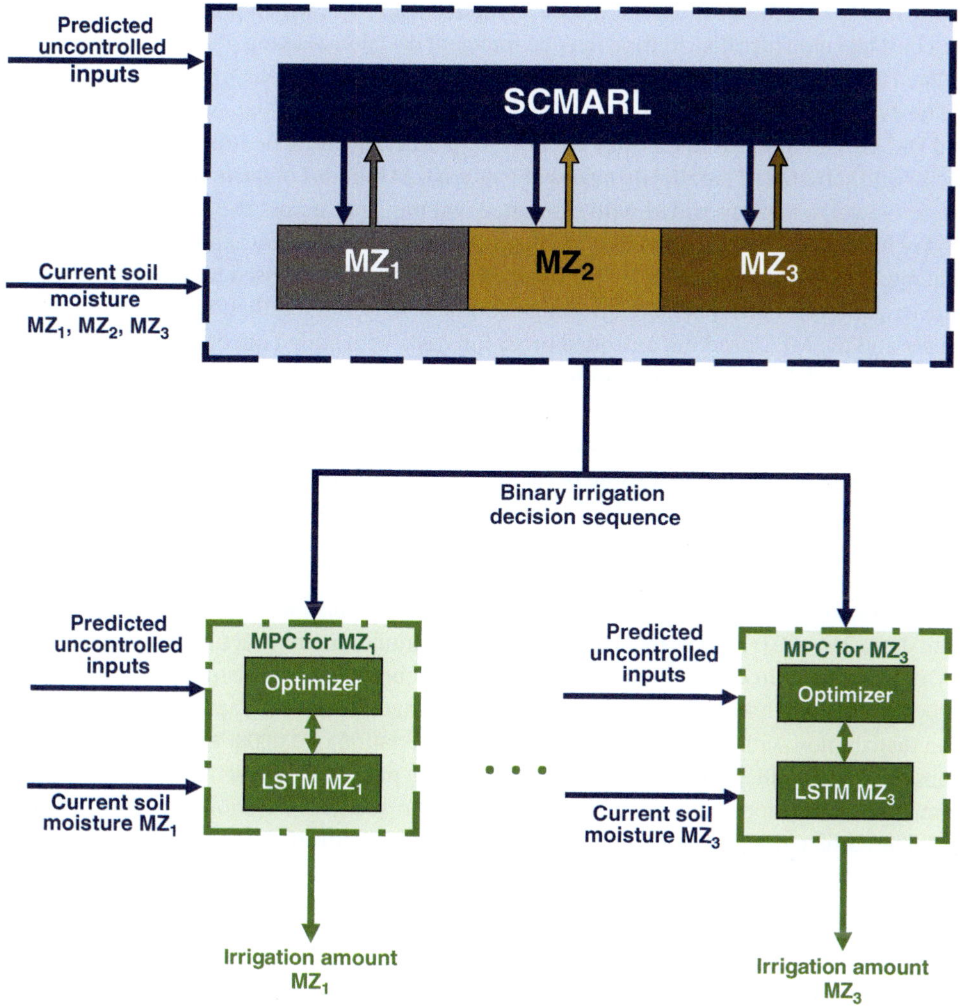


**Figure 8.3** Summary of the SCMARL agents' evaluation process during the field-scale experiment. *Source:* Adapted from [109].

The amounts of irrigation that were finally applied across the MZs were computed using the same set of parallel MPCs—namely, the $M$ instances of the formulation $\mathbb{P}_{\mathrm{N}}(y, d)$—as applied in the DMARL of Chapter 7. During this process, the sequence of "yes/no" irrigation actions produced by the SCMARL agents was imposed as fixed inputs, guiding the MPCs in determining the corresponding irrigation amounts. Moreover, the sequence of irrigation amounts proposed by the local agents in the SCMARL was utilized as initial guesses during the solution of the MPCs.

#### 8.4.3.3 Season-long Simulation Setup

As with the setup in Section 8.4.1.2, the evaluation did not assume direct access to the full soil moisture profile within each MZ. To fill this gap, daily readings were taken from simulated sensors placed in each zone, reporting the mean moisture content in the top 25 cm of soil. These sparse observations were used as inputs to an EKF, which provided estimates of the complete moisture distribution needed for agent decision-making. Further information on the EKF configuration for each zone is available in Appendix D.

The evaluation took place from May 15, 2022 to September 4, 2022. Both the SCMARL and DMARL methods used a 14-day window for prediction and scheduling to execute their agents and solve the $M$ parallel instances of $\mathbb{P}_N(y, d)$. At the start of the investigation period, the frameworks were provided with estimated soil moisture across the MZs, along with a 14-day forecast of the uncontrolled inputs (i.e., crop and weather variables). Each method generated a binary sequence of irrigation decisions, which was passed to the MPC to compute irrigation amounts per zone. Although complete seasonal weather data were available, agents and the MPC used uncontrolled input forecasts with added uncertainty. This uncertainty was modeled as Gaussian noise with a mean of zero and a standard deviation that increased with forecast horizon.

Irrigation quantities computed by the MPC were applied using a receding horizon approach, where only the first value in the computed sequence was implemented, and the remaining values were discarded. The field response was simulated with three independently calibrated 1D Richards' equation models, each corresponding to a specific MZ. Each soil profile extended to a vertical distance of 0.5 m and was divided into 21 uniform layers. To reflect natural variability in field conditions, Gaussian noise with zero mean and a standard deviation of 0.0007 was incorporated into the simulation of the actual field conditions.

On day 2, simulated field conditions were used to obtain mean moisture readings for the upper 25 cm of each MZ. Simulation of sensor error involved drawing samples from a Gaussian distribution with a mean of 0 and a standard deviation of 0.0008. These noise realizations were then added to the moisture readings. The noisy and sparse readings were then combined with prior estimates, using the EKF, to correct the soil moisture state in each MZ. The corrected states, together with the forecasted uncontrolled inputs, formed the basis for generating the "yes/no" irrigation action and the corresponding application amounts. This cycle—comprising daily state updates based on measurements and the calculation of schedules using forecasted uncontrolled inputs—was repeated until the end of the simulation period.

### 8.4.4 Results and Discussion

The findings from the experiments are presented across three key areas. It begins with an evaluation of the state augmentation strategy and its impact on agent performance. The next subsection examines the extent to which agents at the local level choose actions that are in agreement with the action of the coordinator. The section concludes with a comparative analysis of the irrigation prescriptions and the resulting soil moisture trajectories produced by the SCMARL and DMARL approaches during the season-long evaluation.

### 8.4.5 Impact of State Augmentation on Agent Performance

Figure 8.4 illustrates a performance comparison between agents trained using the SCMARL framework with state augmentation and the variant without state augmentation. In Figure 8.4a, the mean scores over 9,800 episodes and 10 training runs are displayed. Agents

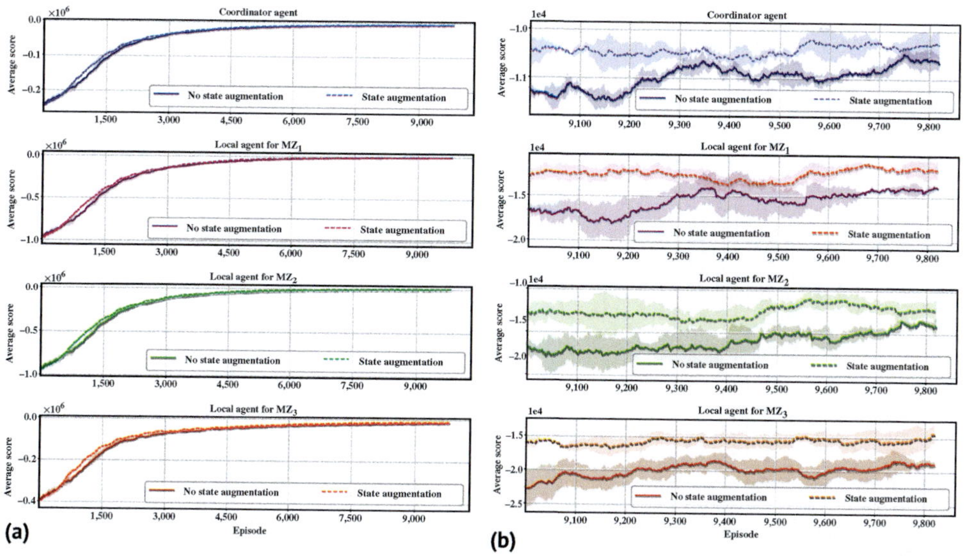


**Figure 8.4** Performance comparison between SCMARL variants with and without state augmentation based on average episode scores. (a) Average scores over 9,800 episodes; (b) average scores over the final 800 episodes. *Source:* Adapted from [109].

operating with state augmentation demonstrate a more stable and higher reward trajectory during the early training phase (up to episode 6,000), with the distinction being most evident in the local agents' performance. To assess the later stages of learning, Figure 8.4b highlights episodes 9,000–9,800. This focused view reveals that agents with state augmentation maintain a higher level of performance during convergence, reflected in consistently elevated average scores relative to those without augmentation.

Figure 8.5 illustrates the soil moisture levels across MZs that resulted from the prescriptions of the SCMARL approaches with and without state augmentation. The figure indicates that both scheduling schemes were effective at achieving the moisture regulation objective throughout the evaluation period. A summary of the computed performance metrics for both approaches is provided in Table 8.1. The results indicate that the SCMARL configuration with state augmentation required 2.3% less irrigation and achieved a 1.9% increase in IWUE relative to the non-augmented version. Combined with its better training performance, these outcomes demonstrate the advantage of integrating state augmentation into the SCMARL framework.

The gains achieved by the version with state augmentation are largely due to the explicit communication of the "yes/no" action to the agents at the local level and the incorporation of this action into the decision-making process of the agents at the local level. This design choice allows local agents to better account for uncertainties in their respective environments, resulting in more stable and effective learning. Similar outcomes have been reported in prior work, such as [110], where the use of hierarchical coordination and directed communication improved robustness in multi-agent learning under non-stationary conditions, particularly in the context of real-time strategy games.

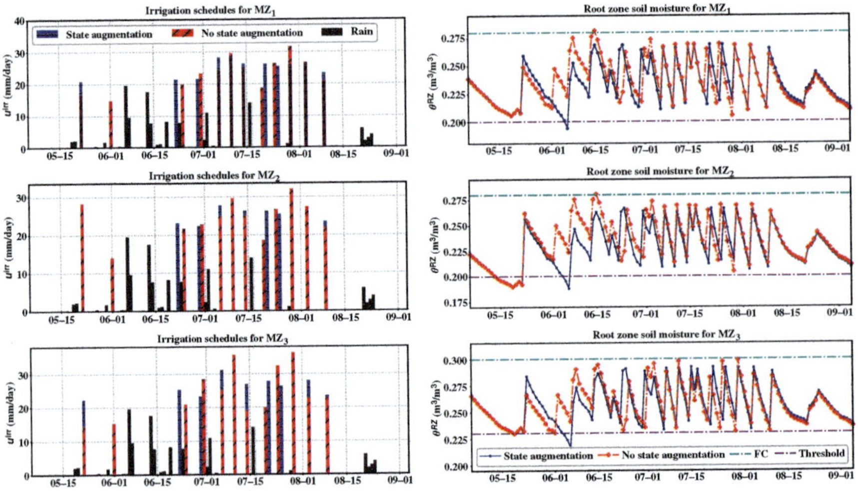


**Figure 8.5** Irrigation prescriptions and corresponding root zone soil moisture content resulting from SCMARL with state augmentation versus without augmentation. *Source:* Adapted from [109].

**Table 8.1** Comparative analysis of SCMARL performance with and without state augmentation.

| Metric | SCMARL (Augmentation) | SCMARL (No augmentation) |
|---|---|---|
| Irrigation for $MZ_1$ | 0.27 | 0.28 |
| Irrigation for $MZ_2$ | 0.28 | 0.29 |
| Irrigation for $MZ_3$ | 0.29 | 0.29 |
| Total irrigation (m) | 0.84 [$\downarrow$ **2.3%**] | 0.86 |
| IWUE (kg/m$^3$) | 3.12 [$\uparrow$ **1.9%**] | 3.06 |

Although the SCMARL approach with state augmentation achieved higher overall performance, the version without state augmentation still performed reliably. Its decentralized training mechanism aligns with independent learning strategies frequently explored in the MARL literature [111, 112]. The learning efficiency observed in this setting can be attributed to two primary factors. First, as discussed in [113], DMARL often functions as an approximation of an independent iterative best response, where each agent adapts its strategy based on the evolving behavior of others. In this SCMARL setting without state augmentation, local agents refine their policies in response to the coordinator's actions. Second, the PPO algorithm's policy clipping mechanism plays a stabilizing role by mitigating certain effects of non-stationarity, as previously shown in [112]. Together, these elements support the framework's ability to learn effectively, even in the absence of explicit coordination signals.

Despite the performance improvements introduced by state augmentation, Table 8.2 shows that the SCMARL framework without state augmentation results in a shorter training duration. This difference is largely due to the reduced computational burden, as the non-augmented version does not involve the additional step of transmitting the coordinator's decision to local agents. Moreover, the resulting policies for local agents are less complex, which reduces the time required for policy updates during training. Consequently, in applications where training efficiency is a priority, the non-augmented SCMARL configuration may be more appropriate. However, as discussed in [113], training agents independently based on iterative best response dynamics may not always yield agent-level optimal solutions. Further investigation is required to determine the conditions under which this configuration can consistently approximate optimal behavior in daily irrigation scheduling scenarios.

### 8.4.6 Local Agents' Policy Agreement with Coordinator's Action

Table 8.3 shows the frequency with which local agents produced actions that aligned with the coordinator agent's decisions, under both the state augmentation and non-augmentation configurations. With state augmentation, alignment was observed in 98% of evaluation cases, compared to just 21% without it. These results indicate that incorporating the coordinator's action into the local agents' input promotes stronger coordination during training. Consequently, agents trained with state augmentation are more likely to act in accordance with global decisions, reducing the need for post-training enforcement mechanisms.

**Table 8.2** Comparison of the total training time required for SCMARL with and without state augmentation, and the DMARL framework.

| | **SCMARL** | **SCMARL** | |
|---|---|---|---|
| **Metric** | **(Augmentation)** | **(No augmentation)** | **DMARL** |
| Time (h) | 9.2 | 8.8 | 1.9 |

**Table 8.3** Evaluation of local agents' alignment with the coordinator agent in the SCMARL framework, with and without state augmentation, based on 2,000 decision-making instances.

| **SCMARL** | **Aligned decisions** | **Misaligned decisions** |
|---|---|---|
| **Framework** | **(Successes)** | **(Failures)** |
| With state augmentation | 1967 (**98%**) | 33 (**2%**) |
| Without state augmentation | 424 (**21%**) | 1576 (**79%**) |

Despite the high alignment rate achieved in the version with state augmentation, misalignment still occurred in 33 of the 2,000 evaluation instances (2%). This suggests that while the approach is effective, some level of inconsistency remains. Prolonging the training period may help further reduce these cases. Furthermore, applying a post-training threshold, such as suppressing irrigation amounts below a threshold value, could help manage the practical effect of occasional misalignments. By contrast, the non-augmented SCMARL framework exhibited an increased misalignment rate of 79%, due to the absence of a mechanism that promotes coordination between local and field-wide decisions.

### 8.4.7 Assessment of SCMARL Performance and Utility

Figure 8.6 shows the soil moisture trajectories across the three MZs under irrigation prescriptions produced by the decentralized and semi-centralized MARL frameworks. Both approaches successfully achieved their moisture regulation objectively and effectively, though infrequent minor boundary deviations were observed across zones. The performance gap between the two methods is illustrated in Table 8.4. The semi-centralized method demonstrated greater resource efficiency, reducing total water use by 3.7% and improving IWUE by 5.6% compared to the decentralized approach.

These differences stem from the distinct coordination strategies employed. The DMARL framework, described in Chapter 7, uses a heuristic mechanism to unify binary irrigation decisions without global optimization, often resulting in less frequent but higher-volume irrigation events. This behavior sometimes leads to moisture levels exceeding the upper threshold of the target range, which can compromise crop growth and reduce water-use efficiency. In contrast, SCMARL's field-wide coordination enables more refined binary

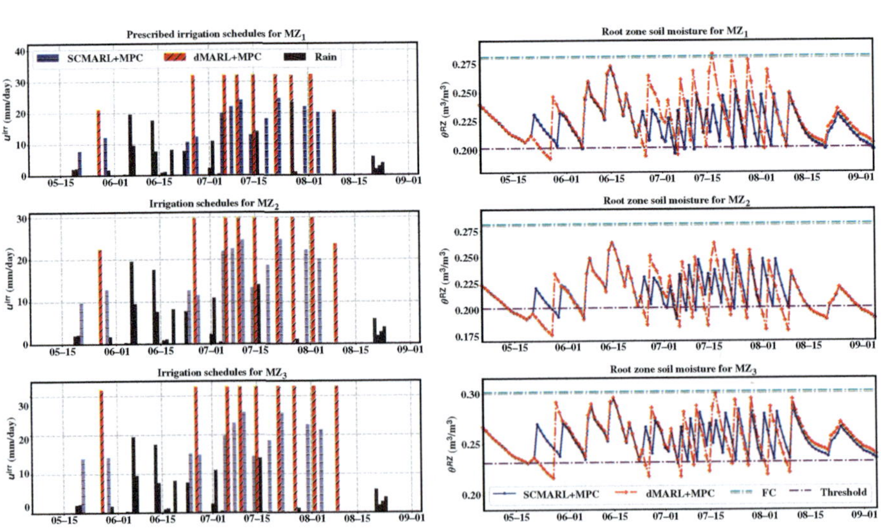


Figure 8.6 Irrigation prescriptions and root zone soil moisture dynamics resulting from the SCMARL and DMARL approaches. *Source:* Adapted from [109].

**Table 8.4** Quantitative comparison between the SCMARL and DMARL scheduling approaches.

| Metric | SCMARL | DMARL |
|---|---|---|
| Irrigation for $MZ_1$ (m) | 0.25 | 0.26 |
| Irrigation for $MZ_2$ (m) | 0.25 | 0.25 |
| Irrigation for $MZ_3$ (m) | 0.27 | 0.29 |
| Total irrigation (m) | 0.77 [↓ **3.7%**] | 0.80 |
| IWUE (kg/m$^3$) | 3.40 [↑ **5.6%**] | 3.22 |

decision-making. When these decisions are integrated into the MPC step, they yield prescriptions that better balance water savings, soil moisture regulation, and optimal crop development.

#### 8.4.7.1 Enhancing SCMARL Performance Through MPC Integration

In contrast to the DMARL method, which relies on a decentralized MPC to compute final irrigation amounts, the SCMARL framework is structured to operate with minimal dependence on such a component. This is made possible by the alignment between the coordinator's decisions and the irrigation amounts generated by local agents. Nevertheless, Tables 8.1 and 8.4 demonstrate that supplementing SCMARL with a decentralized MPC step yields further benefits—achieving an additional 8.3% reduction in total water use and an 8.9% increase in IWUE. The validity of this comparison is ensured by the use of a consistent simulation setup across both approaches, including identical field conditions, estimator configurations, weather data, and noise parameters.

The earlier observation underscores the potential for RL and MPC to function in a complementary fashion—a technique supported by prior studies that have explored the joint application of these methods [114, 115]. In the context of irrigation scheduling, the learning-based multi-agent MPC framework of Chapter 7 represents the first known attempt to integrate RL and MPC in a complementary manner. The results presented in this chapter show that the integration of MPC into the SCMARL framework yields improved performance compared to this earlier approach. This improvement is primarily attributed to the field-wide optimization of daily binary irrigation decisions enabled by the SCMARL design. Moreover, the irrigation amounts generated by SCMARL's local agents are more consistent with the coordinator's binary decisions, making them more suitable as initial conditions for the MPC solver than those provided by the decentralized agents in the learning-based approach of Chapter 7.

This integration proves especially advantageous given the individual limitations of RL and MPC. On its own, SCMARL poses challenges related to computational cost and the complexity of policy design. In particular, modifying the reward function necessitates retraining of the agent network, which can be time-intensive. Conversely, MPC—as implemented in Chapter 6—can be computationally demanding, particularly when tasked with solving for an input space that consists of both continuous and binary components. By combining SCMARL with MPC, the resulting framework relies on the strengths of both: RL contributes adaptability and the ability to learn from dynamic interactions, while MPC brings precision and robustness in adjusting controls.

#### 8.4.7.2 SCMARL Training Costs and Practical Implementation

Although the SCMARL framework outperforms the learning-based multi-agent MPC approach in terms of scheduling efficiency, this advantage comes with higher training costs. On the hardware setup used in this study—an Intel Core i7-8700 CPU (3.20 GHz) with 32 GB of RAM—training the SCMARL agents required 9.2 hours, whereas the DMARL approach completed training in 1.9 hours under identical conditions (see Table 8.2). This highlights a key limitation for settings with limited local computational resources. In such cases, offloading the training process to cloud-based infrastructure can offer a feasible workaround. In contrast, the post-training deployment phase of SCMARL is far less resource-intensive. Once trained, agents can generate irrigation decisions in real time using locally available data. Consequently, these decisions can be calculated with standard farm-level computing systems.

#### 8.4.7.3 Adapting SCMARL to Diverse Agricultural Contexts

Although the SCMARL framework was applied to a specific large-scale field in this study, its design allows for broader applicability to other agricultural contexts. Generalization to a new environment would require recalibrating the underlying 1D Richards' equations using local soil hydraulic characteristics and updating crop-specific parameters and weather data. However, because agent behavior is closely tied to localized environmental dynamics, direct policy transfer between distinct agricultural settings may not be effective. In such cases, retraining the agents would likely be necessary.

To reduce the retraining burden, transfer learning may offer a viable strategy for improving adaptability. Knowledge acquired from one environment can be fine-tuned for use in a similar setting by modifying key parameters such as crop coefficients or soil properties. Future work could explore how transfer learning techniques, such as those surveyed in [116], can support efficient adaptation of SCMARL across varied agricultural settings.

#### 8.4.7.4 Scalable Irrigation Scheduling with SCMARL

The scalability of the SCMARL framework can be assessed from both local and central decision-making perspectives. On the side of the local agents, the decentralized structure naturally supports expansion, as additional MZs can be accommodated without compromising the computational efficiency of the framework. Each local agent makes decisions independently, enabling the framework to scale efficiently across large fields. Training time in such settings can be reduced further through parallelization, which allows simultaneous policy updates and environmental simulations across zones.

However, from the coordinator's perspective, scalability introduces new challenges. As the number of MZs increases, the dimensionality of the coordinator's input also grows, leading to longer computation times during training. A practical alternative is to replace the full moisture distribution with a scalar metric, such as root zone soil moisture content, for each zone. This simplification reduces input complexity and improves computational efficiency, though it may result in a loss of detailed spatial information that could affect decision quality. Future studies could examine the performance impact of this abstraction to determine whether the trade-off is acceptable in large-scale implementations.

## 8.5 Conclusion

This chapter has shown that integrating a hierarchical structure into SCMARL framework—by dividing the irrigation scheduling task into two distinct decision-making layers—advances the application of MARL in managing daily irrigation across spatially heterogeneous fields. The simulation results demonstrated that assigning irrigation timing decisions to a centralized coordinator enhances decision optimality, thereby improving the overall accuracy of the scheduling process. Simultaneously, delegating irrigation amount decisions to decentralized agents takes advantage of the weak interdependencies among MZs, offering a scalable solution for large-scale deployments. Furthermore, the framework's hierarchical design enabled the implementation of a state augmentation strategy, where the centralized coordinator's decisions are integrated into the local agents' input. This mechanism effectively mitigates non-stationarity, contributing to more stable learning across the local agents.

# 9

# Integrating Daily Scheduling and Hourly Control for Precision Irrigation

## 9.1 Introduction

The semi-centralized irrigation scheduling strategy presented in Chapter 8 introduced a hierarchical decision-making structure composed of a centralized coordinating agent and multiple decentralized agents. This architecture provided a scalable and optimal framework for addressing the daily irrigation scheduling problem in large-scale agricultural fields characterized by spatial variability. In comparison to the purely decentralized approach described in Chapter 7, the semi-centralized method was more effective in realizing the full benefits of formulating the daily irrigation scheduling problem as a mixed-integer optimal control problem.

Despite these benefits, the training of agents within the semi-centralized multi-agent reinforcement learning (SCMARL) framework was conducted under a fully observable Markov decision process (MDP) assumption, in which each agent was assumed to have full knowledge of its state. This state comprised the complete soil moisture distribution along with daily weather and crop-related variables. While the weather and crop components can be reasonably obtained from forecasts, the assumption of full access to the soil moisture distribution remains impractical. In most large-scale agricultural fields, soil moisture sensors provide sparse and localized data, making full observability impractical.

To enhance the practical applicability of the SCMARL approach, it becomes necessary to relax the assumption of full observability. A more suitable formulation considers the availability of sparse measurements as a proxy for the true state of soil moisture in different management zones (MZs). This setting is appropriately captured by the partially observable Markov decision process (POMDP) framework, which provides a systematic approach to decision-making under partial observability. In a POMDP, the agent operates in an environment where the true system state is not directly accessible. Actions taken by the agent influence the evolution of the underlying state and yield both observations and rewards. Since the observation does not satisfy the Markov property, the policy must rely on information accumulated over time. This dependence on historical data significantly increases the computational complexity of the learning process, a challenge commonly referred to as the curse of history.

*Precision Irrigation for Agriculture: Integrating Machine Learning and Optimal Control Strategies*, First Edition. Bernard Twum Agyeman and Jinfeng Liu.

The challenge posed by the curse of history is addressed through the introduction of a belief state, which defines a probability distribution over the true system states, conditioned on the sequence of past actions and observations. Belief-state estimation is typically performed through a two-step recursive process: a prediction step, based on the system's transition model, followed by a correction step, informed by the most recent observation. In systems governed by linear dynamics and observation models, belief-state updates can be efficiently performed using the Kalman filter (KF). For nonlinear systems, belief estimation requires more advanced techniques such as the extended Kalman filter (EKF) or particle filtering methods.

As demonstrated in Chapters 4, 5 and 8, the EKF provides a reliable method for estimating soil moisture states from sparse sensor measurements. This is achieved by leveraging a physically based process model (i.e., the Richards' equation) and an observation model (i.e., the van Genuchten soil-water retention relationship). Together, these models offer a sufficiently descriptive and computationally tractable foundation for soil moisture estimation in agro-hydrological systems. Within a POMDP-based SCMARL framework, the EKF facilitates the construction of belief states that represent the underlying soil moisture dynamics under conditions of partial observability.

Up to this point, the focus of this book has been on daily-level irrigation scheduling, where daily inputs such as weather, crop growth, and soil moisture information are used to determine irrigation timing and water application amounts. Simulation experiments have shown that this level of granularity is effective in achieving season-long irrigation efficiency and crop productivity. However, advances in environmental sensing now make it possible to obtain weather data at sub-daily resolutions, particularly at hourly intervals. While daily averages provide a useful summary, they fail to capture the variations that occur in weather variables throughout the day. These intraday variations can influence soil moisture dynamics and, consequently, affect the irrigation requirements of crops.

Incorporating hourly variability into the irrigation decision-making process can significantly enhance both responsiveness and control precision. However, for a mixed-integer optimal control-based scheduler, the direct integration of hourly weather data alongside daily data would substantially increase the computational complexity of an already demanding problem. To manage this issue effectively, a hierarchical framework, similar to those used in process systems engineering, can be adopted. Such frameworks typically decompose decision-making across multiple temporal layers, including planning, scheduling, real-time optimization, advanced control, and regulatory control, each tailored to a specific timescale and associated objective.

For irrigation applications, the scheduling and advanced control layers offer a computationally efficient means of leveraging multi-timescale weather information in decision-making. These layers can be integrated in a hierarchical manner to efficiently utilize both daily and hourly information. The scheduling layer, operating on a daily timescale, determines whether irrigation should be carried out and prescribes the total water volume to be applied per MZ. These directives are then passed to the advanced control layer, which operates at an hourly resolution. Using model predictive control (MPC), this layer dynamically adjusts irrigation amounts throughout the day to track the prescribed daily schedules while incorporating updated soil moisture estimates and high-resolution weather forecasts.

Based on the foregoing, this chapter presents a hierarchical framework that integrates an SCMARL-based scheduler trained within a POMDP setting with an advanced control layer. In this learning framework of the SCMARL, belief states are constructed using the EKF. Positioned at the upper level of the hierarchy, the SCMARL scheduler leverages daily soil moisture estimates, along with daily weather and crop information, to generate daily irrigation schedules. At the lower level, an MPC-based advanced control layer operates using hourly inputs (i.e., weather forecasts, crop conditions, and sensor feedback) to track the high-level schedules. The proposed hierarchical framework is applied to a large-scale field scenario for performance evaluation.

## 9.2 Observer-based SCMARL Scheduler

As previously noted, scheduling in the upper layer of the hierarchical framework is carried out using an observer-based SCMARL approach. This structure builds on the framework introduced in Chapter 8, which consists of a coordinator agent that aggregates field-wide information to determine whether irrigation should be initiated on a given day. The coordinator's decision is then communicated to a layer of decentralized local agents. Each local agent integrates this high-level directive with its own localized data to generate zone-specific irrigation schedules.

The key advancement introduced in this chapter lies in the formulation of the SCMARL framework within a POMDP setting. Under this formulation, belief states are constructed using the EKF. These belief states serve as the basis for policy optimization across both the coordinator and local agents.

This section first provides a formal treatment of the POMDP framework. It then details the application of the EKF for belief-state estimation under sparse sensing. Finally, a modified SCMARL algorithm is presented that explicitly integrates belief-state estimation into the learning process.

### 9.2.1 POMDP Setting

A POMDP is defined as a tuple $\langle \mathcal{S}, \mathcal{A}, \mathcal{T}, \mathcal{R}, \Omega, \mathcal{O} \rangle$ [117], where:

- $\mathcal{S}$ is a finite set of states;
- $\mathcal{A}$ is a finite set of actions;
- $\mathcal{T} : \mathcal{S} \times \mathcal{A} \times \mathcal{S} \to [0,1]$ is the state transition function, where $\mathcal{T}(s,a,s') = \Pr(s' \mid s,a)$ denotes the probability of transitioning to state $s'$ given the current state $s$ and action $a$;
- $\mathcal{R} : \mathcal{S} \times \mathcal{A} \to \mathbb{R}$ is the reward function, specifying the expected immediate reward received for taking action $a$ in state $s$;
- $\Omega$ is a finite set of observations;
- $\mathcal{O} : \mathcal{A} \times \mathcal{S} \times \Omega \to [0,1]$ is the observation function, where $\mathcal{O}(a,s',o) = \Pr(o \mid a,s')$ denotes the probability of receiving observation $o$ after taking action $a$ and transitioning to state $s'$.

In a POMDP, the agent does not directly observe the true state of the environment. Instead, it receives a sequence of observations that provide partial information about the underlying state. While certain observation histories may allow the agent to infer the exact state, in general, uncertainty remains. This uncertainty is captured by a *belief state*, denoted by $b$, which represents a probability distribution over all possible states, conditioned on the agent's history of actions and observations. The belief state is sufficient in the sense that the agent can learn the same policies as if it had access to history.

Following the execution of action $a$ and receipt of observation $o$, the belief state is updated to $b'$ according to the following rule:

$$b'(s') = \frac{\mathcal{O}(a, s', o)\sum_{s \in S} \mathcal{T}(s, a, s')b(s)}{\Pr(o \mid b, a)} \tag{9.1}$$

where $\Pr(o \mid b, a)$ is a normalizing constant ensuring that $b'$ is a valid probability distribution. This recursive update mechanism enables the agent to maintain and refine an internal estimate of the unobservable environmental state, which is essential for decision-making.

In principle, any POMDP can be reformulated as a fully observable MDP over the space of belief states, commonly referred to as the belief-state MDP. In this formulation, the belief state serves as the observable state in the reformulated MDP.

The reward function in the belief-state MDP is defined as the expected reward with respect to the belief. Specifically, if $R(x, a)$ denotes the immediate reward for taking action $a$ in state $x$, then the corresponding reward function over belief states is given by:

$$r(b, a) = \sum_{x \in S} b(x)R(x, a), \tag{9.2}$$

which represents the expected reward for executing action $a$ in belief state $b$. This formulation enables the use of standard MDP solution methods over the continuous space of belief states.

### 9.2.2 EKF for Belief-state Estimation

Equation (9.1) represents a Bayes filter update, which forms the basis of the KF [118]. Accordingly, belief-state estimation can be achieved using the KF in linear systems or the EKF in systems governed by nonlinear dynamics. The KF operates recursively, requiring only the previous state estimate, the applied control action, and the current observation to produce a new state estimate. In this formulation, the belief state is represented by the mean (the most probable state estimate) and a covariance matrix that quantifies the associated uncertainty.

As demonstrated in Chapters 4, 5, and 8, the EKF provides accurate estimates of soil moisture in agro-hydrological systems. It thus serves as a robust estimator for belief-state construction within the POMDP formulation of the SCMARL framework. Recall that the agents in SCMARL are trained independently for each MZ, leveraging the weak coupling between zones in large-scale fields. This independence allows for belief-state estimation on a per-zone basis, with an EKF tailored to each MZ.

For a given MZ, the EKF assimilates noisy soil moisture observations—resulting from the control actions of its corresponding local agent—into a physics-based model. The physics-based model employed is a calibrated one-dimensional (1D) Richards' equation, tailored to the soil characteristics of each MZ. The EKF produces both an estimate of the belief state (i.e., the soil moisture profile) and its associated covariance.

To implement the EKF, both the process (transition) and observation models must be specified. Here, the transition model is the 1D Richards' equation calibrated using hydraulic parameters representative of the MZ soil properties. The observation model is based on the van Genuchten relation, which links volumetric water content to capillary pressure head:

$$C(\psi)\frac{\partial \psi}{\partial t} = \frac{\partial}{\partial z}\left[K(\psi)\frac{\partial(\psi + z)}{\partial z}\right] \tag{9.3}$$

$$\theta_v(\psi) = \theta_r + (\theta_s - \theta_r)\left[\frac{1}{1 + (-\alpha\psi)^n}\right]^{1-\frac{1}{n}} \tag{9.4}$$

For completeness, the reader is referred to Equations (2.21) and (2.9) in Chapter 2 for detailed definitions. These models are reformulated into a discrete-time nonlinear state-space form:

$$x_{k+1} = \mathcal{F}(x_k, u_k^{\text{irr}}, \text{ET}_{0_k}, R_{n_k}, K_{c_k}, \boldsymbol{\phi}) + \omega_k \tag{9.5}$$

$$y_k = \mathcal{H}(x_k, \boldsymbol{\phi}) + v_k \tag{9.6}$$

where $x_k \in \mathbb{R}^{N_x}$ denotes the capillary pressure head at $N_x$ spatial points, $u_k^{\text{irr}}$ is the irrigation amount, $\text{ET}_{0_k}$ is the reference evapotranspiration, $K_{c_k}$ is the crop coefficient, and $R_{n_k}$ is the rainfall. The parameter vector $\boldsymbol{\phi}$ captures soil hydraulic properties, while $\omega_k$ and $v_k$ are process and measurement noise terms, respectively. The observation $y_k$ consists of predicted volumetric water content at the same spatial locations.

The EKF proceeds in the following steps:

- **Initialization:** Initialize the state estimate $\hat{x}_0^+$, error covariance $P_0^+$, process noise covariance $Q$, and measurement noise covariance $R$.
- **Prediction:** For time $k + 1$,

$$\hat{x}_{k+1}^- = \mathcal{F}(\hat{x}_k^+, u_k, \boldsymbol{\phi}) \tag{9.7}$$

$$P_{k+1}^- = A_k P_k^+ A_k^\top + Q \tag{9.8}$$

  where $A_k = \left.\frac{\partial \mathcal{F}}{\partial x}\right|_{\hat{x}_k^+, u_k}$ is the Jacobian of the process model and $u_k$ is a compact representation of the inputs $u_k^{\text{irr}}, \text{ET}_{0_k}, R_{n_k}$, and $K_{c_k}$.
- **Update:** Given the observation $o_{k+1}$ at time $k + 1$,

$$\hat{y}_{k+1}^- = \mathcal{H}(\hat{x}_{k+1}^-, \boldsymbol{\phi}) \tag{9.9}$$

$$C_{k+1} = \left.\frac{\partial \mathcal{H}}{\partial x}\right|_{\hat{x}_{k+1}^-} \tag{9.10}$$

$$G_{k+1} = P_{k+1}^- C_{k+1}^\top (C_{k+1} P_{k+1}^- C_{k+1}^\top + R)^{-1} \tag{9.11}$$

$$\hat{x}_{k+1}^+ = \hat{x}_{k+1}^- + G_{k+1}(o_{k+1} - \mathcal{M}\hat{y}_{k+1}^-) \tag{9.12}$$

$$P^{+}_{k+1} = (I - G_{k+1}C_{k+1})P^{-}_{k+1} \tag{9.13}$$

where $\mathcal{M}$ is a selection matrix that maps the predicted soil moisture profile to the observation $o$ and $C_{k+1}$ is the Jacobian of the observation function. The Kalman gain $G_{k+1}$ balances model predictions and sensor feedback. The updated output estimate is:

$$\hat{y}^{+}_{k+1} = \mathcal{H}(\hat{x}^{+}_{k+1}, \phi) \tag{9.14}$$

Recall that in the SCMARL of Chapter 8, the soil moisture profile $y$, and not the capillary head profile $x$, serves as input to the agents of the framework. Therefore, at each time step $k+1$, the belief state, expressed in terms of soil moisture, is represented by the posterior estimate $\hat{y}^{+}_{k+1}$, computed with Equation (9.14) from $\hat{x}^{+}_{k+1}$.

### 9.2.3 Belief State-based Reward Functions

In the original SCMARL framework, agent rewards were computed using the true root zone soil moisture distribution $y$. However, within a POMDP formulation, $y$ is unobservable and must be estimated from noisy observations $o$ using the EKF, yielding an estimate $\hat{y}$. Consequently, prior to presenting the observer-based SCMARL algorithm, the reward functions of both the local and coordinator agents are reformulated to depend on belief-based estimates.

#### 9.2.3.1 Local Agents

For a local agent, the belief-based reward $\hat{r}_{\text{la}}$ consists of two components: a soil moisture regulation term $\hat{r}^{z}_{\text{la}}$ and an irrigation usage minimization term $r^{u}_{\text{la}}$:

$$\hat{r}_{\text{la}} := \alpha_{\text{la}}\hat{r}^{z}_{\text{la}} + \beta_{\text{la}}r^{u}_{\text{la}} \tag{9.15}$$

where $\alpha_{\text{la}}$ and $\beta_{\text{la}}$ reflect the trade-off between maintaining soil moisture within the target range and minimizing water use. The term $\hat{r}^{z}_{\text{la}}$ evaluates how well the agent maintains the root zone soil moisture estimate, $\hat{\theta}^{\text{RZ}}$, within the prescribed soil moisture range $[\underline{\nu}, \overline{\nu}]$. This estimate is derived from the belief state $\hat{y}^{+}$ using the weighted method detailed Section 7.3.2 of Chapter 7:

$$\hat{r}^{z}_{\text{la}} := \begin{cases} -\underline{Q} \cdot |\hat{\theta}^{\text{RZ}} - \underline{\nu}| & \text{if } \hat{\theta}^{\text{RZ}} < \underline{\nu} \\ -\overline{Q} \cdot |\hat{\theta}^{\text{RZ}} - \overline{\nu}| & \text{if } \hat{\theta}^{\text{RZ}} > \overline{\nu} \\ 0 & \text{if } \underline{\nu} \leq \hat{\theta}^{\text{RZ}} \leq \overline{\nu} \end{cases} \tag{9.16}$$

Here, $\underline{Q} > 0$ and $\overline{Q} > 0$ are penalty weights that emphasize deviations below or above the target moisture range, respectively.

The component of the reward that penalizes irrigation usage, $r^{u}_{\text{la}}$, is expressed as:

$$r^{u}_{\text{la}} := -R_u a_{\text{la}} \tag{9.17}$$

where $R_u > 0$ denotes the cost assigned to each unit of water applied.

#### 9.2.3.2 Coordinator Agent

The coordinator agent's belief-based reward $\hat{r}_{\mathrm{ca}}$ is also composed of two components: an aggregated moisture regulation reward $\hat{r}^{z}_{\mathrm{ca}}$ and a cost minimization term $r^{c}_{\mathrm{ca}}$:

$$\hat{r}_{\mathrm{ca}} := \alpha_{\mathrm{ca}}\hat{r}^{z}_{\mathrm{ca}} + \beta_{\mathrm{ca}}r^{c}_{\mathrm{ca}} \tag{9.18}$$

where $\alpha_{\mathrm{ca}}$ and $\beta_{\mathrm{ca}}$ denote the respective weighting parameters. The moisture regulation component is computed by summing the local moisture regulation rewards across all $M$ MZs:

$$\hat{r}^{z}_{\mathrm{ca}} := \sum_{i=1}^{M} \hat{r}^{z}_{\mathrm{la},i} \tag{9.19}$$

with $\hat{r}^{z}_{\mathrm{la},i}$ given by Equation (9.16).

The cost minimization term penalizes the execution of the irrigation command:

$$r^{c}_{\mathrm{ca}} := -R_c c \tag{9.20}$$

where $R_c > 0$ specifies the cost associated with initiating irrigation, and $c \in \{0, 1\}$ denotes the daily binary irrigation decision.

The complete training procedure of the observer-based SCMARL framework, which integrates EKF-based belief-state estimation and modified reward functions within the POMDP formulation, is presented in Algorithm 4. For a comprehensive explanation of the SCMARL framework, readers are referred to Chapter 8.

## 9.3 Advanced Controller Design

As previously mentioned, the advanced controller operates beneath the scheduling layer and is tasked with implementing the daily irrigation prescriptions while accounting for hourly variations in both weather and soil moisture. This controller employs an MPC strategy, implemented using a shrinking horizon approach. The adoption of a shrinking rather than fixed horizon is driven by several important considerations. More importantly, the shrinking horizon ensures that the MPC operates strictly within the temporal scope of the scheduler's daily prescription. A fixed horizon, when paired with receding horizon control, would require extending forecasts and control decisions beyond the current day, potentially violating the scheduler's prescribed daily irrigation decision. Furthermore, the shrinking horizon guarantees that the controller focuses only on time periods where irrigation can actually be applied, thereby avoiding evaluations that include hours beyond the controller's jurisdiction. In this way, the MPC can implement receding horizon control in a manner that remains fully aligned with the scheduler's decisions.

The MPC is activated only on days when the scheduler prescribes irrigation. Since these prescriptions by the scheduler are made on a per-MZ basis, a dedicated MPC is assigned to each MZ within the field. For a given MZ, the MPC is initialized with the current estimate of the soil moisture profile and a 24-hour weather forecast for the day. The controller employs a 1D Richards' equation model of the soil in the MZ to compute hourly irrigation amounts that collectively fulfill the daily irrigation amount specified by the scheduler, while also driving the root zone soil moisture toward the target value associated with that irrigation amount.

**Algorithm 4** Pseudocode of the Observer-based SCMARL Algorithm

Initialize weights of critic $C_{ca}$ and actor $\mathcal{A}_{ca}$ networks of the coordinator.
**for** local agent $i \leftarrow 1$ to $M$ **do**
  Initialize weights of critic $C_{la_i}$ networks of $i$.
  Initialize weights of actor $\mathcal{A}_{la_i}$ networks of $i$.
**end for**
Set the total number of episodes $K$.
Set the horizon length for each episode $T$.
**for** episode $k \leftarrow 1$ to $K$ **do**
  **for** local agent $i \leftarrow 1$ to $M$ **do**
    Set $\hat{x}_0^+$, $P_0^+$, $Q$, and $R$ of $i$.
    Calculate $(\hat{y}_{i,\text{init}})$ of $i$ using $\hat{x}_0^+$ according to Equation (9.14).
    Set $\hat{y}_i \leftarrow \hat{y}_{i,\text{init}}$.
    Set $\hat{x}_i^+ \leftarrow \hat{x}_0^+$ and $P^+ \leftarrow P_0^+$.
  **end for**
  Obtain initial weather and crop coefficient data $\text{ET}_{0,\text{init}}, R_{n,\text{int}}, K_{c,\text{int}}$.
  Set $\text{ET}_0 \leftarrow \text{ET}_{0,\text{init}}, R_n \leftarrow R_{n,\text{int}}, K_c \leftarrow K_{c,\text{int}}$.
  **for** timestep $t \leftarrow 1$ to $T$ **do**
    Obtain $s_{ca} = [\hat{y}_1, \dots, \hat{y}_M, \text{ET}_0, K_c, R_n]$.
    Select $a_{ca} = c = \mathcal{A}_{ca}(s_{ca})$.
    **for** local agent $i \leftarrow 1$ to $M$ **do**
      Obtain $s_{la_i} = [\hat{y}_i, \text{ET}_0, K_c, R_n, c]$.
      Obtain $a_{la_i} = \mathcal{A}_{la_i}(s_{la_i})$ for agent $i$.
      Execute $u_{la_i}^{\text{irr}} = a_{la_i} \times c$ in environment of $i$.
      Obtain observation $o_i$ from the environment of $i$.
      Update $\hat{x}_i^+$ & $P^+$ with Equations (9.7)–(9.8) to get $\hat{x}_i^-$ & $P_i^-$.
      Update $\hat{x}_i^-$ & $P_i^-$ with Equations (9.9)–(9.13) to get $\hat{x}_i^+$ & $P^+$.
      Compute $\hat{y}^+$ using $\hat{x}^+$ according to Equation (9.14).
      Obtain $\hat{r}_{la_i}$ from Equation (9.15).
    **end for**
    Store $[s_{la_1}, \dots, s_{la_M}]$, $[a_{la_1}, \dots, a_{la_M}]$, $[\hat{r}_{la_1}, \dots, \hat{r}_{la_M}]$, and $[\hat{y}_{la_1}^+, \dots, \hat{y}_{la_M}^+]$.
    Obtain weather and crop coefficient predictions as $\text{ET}_0^+, R_n^+, K_c^+$.
    Obtain $s_{ca}^+ = [\hat{y}_1^+, \hat{y}_2^+, \dots, \hat{y}_M^+, \text{ET}_0^+, K_c^+, R_n^+]$.
    Obtain $\hat{r}_{ca}$ from Equation (9.18).
    Store $(s_{ca}, a_{ca}, \hat{r}_{ca}, s_{ca}^+)$ for coordinator.
    Update $C_{ca}$ and $\mathcal{A}_{ca}$.
    Obtain $c^+ = \mathcal{A}_{ca}(s_{ca}^+)$.
    **for** local agent $i \leftarrow 1$ to $M$ **do**
      Obtain $s_{la_i}^+ = [\hat{y}_i^+, \text{ET}_0^+, K_c^+, R_n^+, c^+]$.
      Store $(s_{la_i}, a_{la_i}, \hat{r}_{la_i}, s_{la_i}^+)$ for $i$.
      Update $C_{la_i}$ and $\mathcal{A}_{la_i}$ for $i$.
    **end for**
    **for** local agent $i \leftarrow 1$ to $M$ **do**
      Set $\hat{y}_i \leftarrow \hat{y}_i^+$.
    **end for**
    Set $\text{ET}_0 \leftarrow \text{ET}_0^+, R_n \leftarrow R_n^+, K_c \leftarrow K_c^+$.
  **end for**
**end for**

After solving the MPC optimization problem, only the irrigation amount for the first hour is applied. This application causes a state transition in the soil moisture profile. The applied amount is subtracted from the remaining daily irrigation prescription, and the updated soil state—alongside revised weather forecasts for the remaining 23 hours—is used to resolve the MPC. This rolling process continues for each subsequent hour, with the prediction and control horizons shrinking by one step at every iteration.

By the end of the 24-hour cycle, the MPC will have sequentially adjusted its hourly control actions in response to updated soil moisture estimates and hourly weather forecasts, thereby seeking to track both the scheduler-prescribed daily irrigation amount and the corresponding root zone soil moisture reference. Notably, these targets are not enforced as hard constraints within the controller design. Rather, they are incorporated into the MPC cost function, which permits deviations when necessary. However, such deviations are penalized to encourage adherence to the scheduler's directives. This design strikes a balance between responsiveness to real-time conditions and adherence to the scheduler's prescriptions.

The MPC problem $\mathbb{P}(h, \{\mathrm{W}\}_h^{h+N_c-1}, N_c, u_{\mathrm{ref}}^{\mathrm{irr}}, \theta_{\mathrm{ref}}^{\mathrm{RZ}}, \hat{x}_h)$ is formulated for the current hour $h$, using the predicted weather ($\mathrm{ET}_0$ and $R_n$) and crop information ($K_c$) from hour $h$ to hour $h+N_c-1$, which is compactly represented as $\{\mathrm{W}\}_h^{h+N_c-1}$, the estimated capillary pressure head distribution $\hat{x}_h$, control horizon $N_c$, prescribed daily irrigation amount $u_{\mathrm{ref}}^{\mathrm{irr}}$, and the target root zone soil moisture $\theta_{\mathrm{ref}}^{\mathrm{RZ}}$, as follows:

$$\min_{\boldsymbol{u}^{\mathrm{irr}}} \lambda_u \left( \sum_{k=h}^{h+N_c-1} u_k^{\mathrm{irr}} - u_{\mathrm{ref}}^{\mathrm{irr}} \right)^2 + \lambda_\theta \left( \theta_{h+N_c}^{\mathrm{RZ}} - \theta_{\mathrm{ref}}^{\mathrm{RZ}} \right)^2 \tag{9.21a}$$

s.t.

$$x_h = \hat{x}_h \tag{9.21b}$$

$$x_{k+1} = \mathcal{F}\left(x_k, u_k^{\mathrm{irr}}, \mathrm{ET}_{0_k}, R_{n_k}, K_{c_k}, \phi\right), \quad \forall k \in [h, h+N_c-1] \tag{9.21c}$$

$$y_k = \mathcal{H}\left(x_k, \phi\right), \quad \forall k \in [h, h+N_c] \tag{9.21d}$$

$$u_{\min}^{\mathrm{irr}} \le u_k^{\mathrm{irr}} \le u_{\max}^{\mathrm{irr}}, \quad \forall k \in [h, h+N_c-1] \tag{9.21e}$$

$$\theta_k^{\mathrm{RZ}} = \mathcal{R}_z(y_k), \quad \forall k \in [h, h+N_c] \tag{9.21f}$$

where $\boldsymbol{u}^{\mathrm{irr}} := [u_h^{\mathrm{irr}}, \ldots, u_{h+N_c-1}^{\mathrm{irr}}]$. The objective function (Equation 9.21a) considers two key objectives:

- **Objective 1**: The deviation between the total irrigation applied over the horizon and the scheduler-prescribed daily amount $u_{\mathrm{ref}}^{\mathrm{irr}}$.
- **Objective 2**: The deviation of the predicted root zone soil moisture at the end of the horizon, $\theta_{h+N_c}^{\mathrm{RZ}}$, from the target value $\theta_{\mathrm{ref}}^{\mathrm{RZ}}$.

The weights $\lambda_u$ and $\lambda_\theta$ determine the relative importance of Objectives 1 and 2, respectively. The state evolution (Equation 9.21c) begins with the current estimate of the capillary pressure head, $\hat{x}_h$ (Equation 9.21b), and follows the nonlinear soil moisture dynamics represented by $\mathcal{F}(\cdot)$. At each time step, the soil state $x_k$ is transformed into a moisture profile $y_k$ through the output map $\mathcal{H}(\cdot)$ (Equation 9.21d). The corresponding root zone volumetric water content $\theta_k^{\mathrm{RZ}}$ is then computed using the aggregation function $\mathcal{R}_z(\cdot)$ (Equation 9.21f), which compactly encodes the weighted method introduced in Section 7.3.2 of Chapter 7. Control inputs are constrained within permissible bounds through Equation (9.21e).

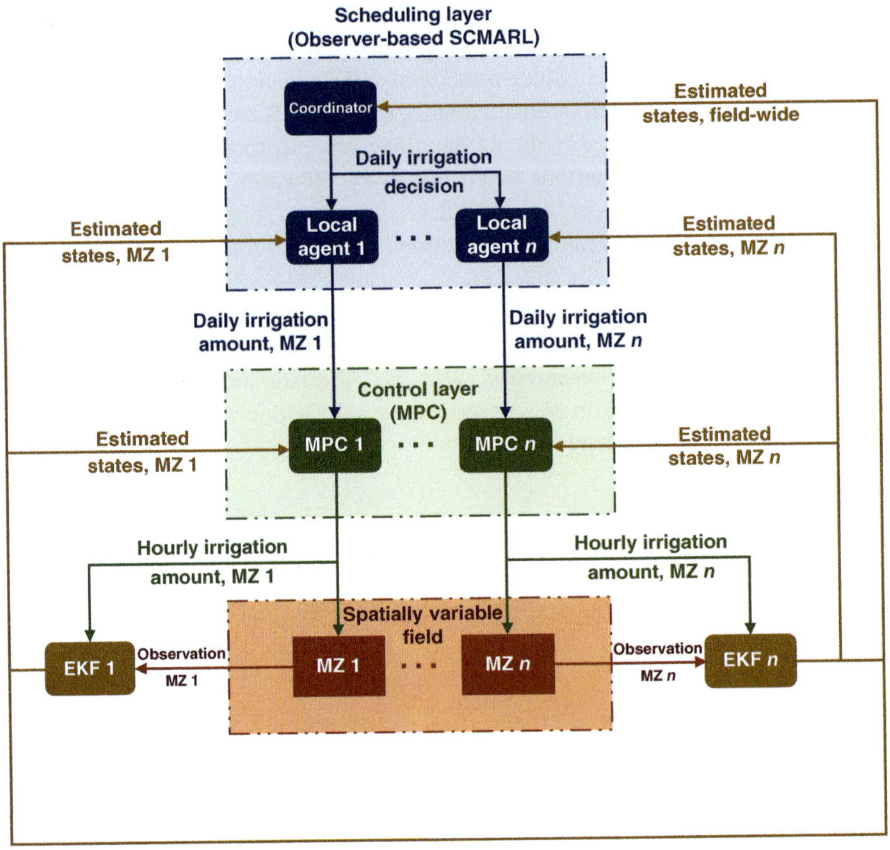


**Figure 9.1** Schematic illustration of the integrated scheduling and control framework, showing the interaction between the SCMARL-based scheduler and the MPC-based control layer.

The procedure for implementing $\mathbb{P}(h, \{W\}_h^{h+N_c-1}, N_c, u_{\text{ref}}^{\text{irr}}, \theta_{\text{ref}}^{\text{RZ}}, \hat{x}_h)$ in a shrinking horizon manner is detailed in Algorithm 5. Additionally, the hierarchical integration of the scheduling and control layers is schematically illustrated in Figure 9.1.

## 9.4 Observer-based SCMARL Implementation

Quadrant 3 of the study area, initially presented in Chapter 7, was selected for training and implementation of the observer-based SCMARL framework and its integration with the MPC controller. This quadrant was partitioned into three MZs using a delineation approach informed by spatial variation in elevation and estimated soil hydraulic parameters. Wheat was uniformly cultivated across all zones throughout the training and implementation period. The SCMARL architecture consisted of a single coordinator agent responsible for irrigation control at the quadrant level and three local agents, each dedicated to managing irrigation within a specific MZ.

**Algorithm 5** Shrinking Horizon MPC for Hourly Irrigation Control

**Require:** Daily prescriptions ($u^{\text{irr}}_{\text{ref}}$ and $\theta^{\text{RZ}}_{\text{ref}}$) and initial pressure estimate $\hat{x}(0)$

**if** $u^{\text{ref}} > 0$ **then**
    Set $\hat{x}_h \leftarrow \hat{x}(0)$
    Set $N_c \leftarrow 24$
    Initialize remaining irrigation: $u^{\text{irr}}_{\text{ref}} \leftarrow u^{\text{irr}}_{\text{ref}}$
    **for** $h \leftarrow 0$ to 23 **do**
        Obtain $\mathrm{W}_h^{h+N_c-1} := (\mathrm{ET}_{0_h}, R_{n_h}, K_{c_h})$ from $h$ to $h + N_c - 1$
        Solve $\mathbb{P}(h, \{\mathrm{W}\}_h^{h+N_c-1}, N_c, u^{\text{irr}}_{\text{ref}}, \theta^{\text{RZ}}_{\text{ref}}, \hat{x}_h)$ to get $\boldsymbol{u}^{\text{irr}} := \{u^{\text{irr}}_k\}_{k=h}^{h+N_c-1}$
        Apply $u^{\text{irr}}_h$
        Obtain the soil moisture observation $o$ and estimate $\hat{x}$
        Update remaining irrigation: $u^{\text{irr}}_{\text{ref}} \leftarrow u^{\text{irr}}_{\text{ref}} - u^{\text{irr}}_h$
        $\hat{x}_h \leftarrow \hat{x}$
        Shrink control horizon: $N_c \leftarrow N_c - 1$
    **end for**
**else**
    Set $u^{\text{irr}}_h \leftarrow 0$ for all $h$
**end if**

### 9.4.1 Environment Configuration and Training

The soil moisture dynamics in each MZ were simulated using a calibrated 1D Richards' equation, with parameters determined from the hydraulic properties at the centroid of each zone. Each MZ was modeled as a 0.50-m vertical soil column discretized into 21 equally spaced layers. Simulations covered a 24-hour period with a fixed time step of 30 minutes. The initial moisture conditions were based on the final profiles from the earlier estimation of soil moisture and hydraulic parameters, which was completed prior to MZ delineation. These conditions were treated as the ground truth for each MZ. The true states were then propagated forward in time, under the action of the agents, to generate soil moisture observations, which were subsequently used to estimate belief states under the POMDP formulation. During this propagation, process noise sampled from a normal distribution with zero mean and a standard deviation of 0.0002 was added to capture process uncertainty.

Each MZ was equipped, in simulation, with a remote sensor assumed to provide daily observations corresponding to the average soil moisture content in the top 25 cm of the soil profile. To account for sensor uncertainty, Gaussian noise with zero mean and a standard deviation of 0.0005 was added to the observed values. For the EKF implementation in each zone, the filter was initialized with a randomly generated initial state estimate $\hat{x}_0^+$, an initial error covariance matrix $P_0^+ = 15.9\mathbb{I}_{21}$, process noise covariance $Q = 0.05\mathbb{I}_{21}$, and measurement noise covariance $R = 19.25$.

Environmental conditions used during training were generated from historical weather records. Daily evapotranspiration values were sampled uniformly from the range 1.04–9.0 mm/day. Rainfall and temperature data were drawn from a dataset covering 2,448 days, corresponding to the wheat-growing seasons from 2005 to 2020. Crop coefficients for wheat were estimated using a temperature-based empirical model (refer to Equation (A.3) in Appendix A.4). A fixed rooting depth of 0.50 m was applied across all MZs.

For the local agents, the reward weighting parameters were set as $\alpha_{\text{la}} = 1.0$ and $\beta_{\text{la}} = 1.0$, while the coordinator agent used $\alpha_{\text{ca}} = 0.1$ and $\beta_{\text{ca}} = 1.0$. The parameters $\bar{Q}$ and $\underline{Q}$, were fixed at 1,200,000 and 1,000,000, respectively. Additional penalty terms were defined as $R_c = 1,000$ and $R_u = 9,000$. A management allowable depletion of 50% was used to compute the soil moisture thresholds. For $MZ_1$ and $MZ_2$, the upper and lower bounds were $\bar{\nu} = 0.280$ and $\underline{\nu} = 0.200$, while for $MZ_3$, the values were $\bar{\nu} = 0.300$ and $\underline{\nu} = 0.230$.

The actor and critic networks were trained with a learning rate of $1 \times 10^{-5}$. All input features were normalized to the range $[-2.0, 2.0]$ using min-max scaling. Each training episode lasted 30 days. Training was conducted using a minibatch size of 64, across 20 epochs, with a discount factor of 0.99, a generalized advantage estimation coefficient of 0.97, a clipping threshold of 0.25, and an entropy regularization term of 0.01. The agents were implemented using the Tensorforce library [107] and trained over 10,000 episodes, across 10 independent runs.

### 9.4.2 Learning Outcomes of Observer-based SCMARL Agents

Figure 9.2 illustrates the evolution of agent performance throughout the training process. The trajectories depict the average score across 10 independent training runs, each comprising 10,000 episodes. During the initial phase (approximately the first 5,000 episodes),

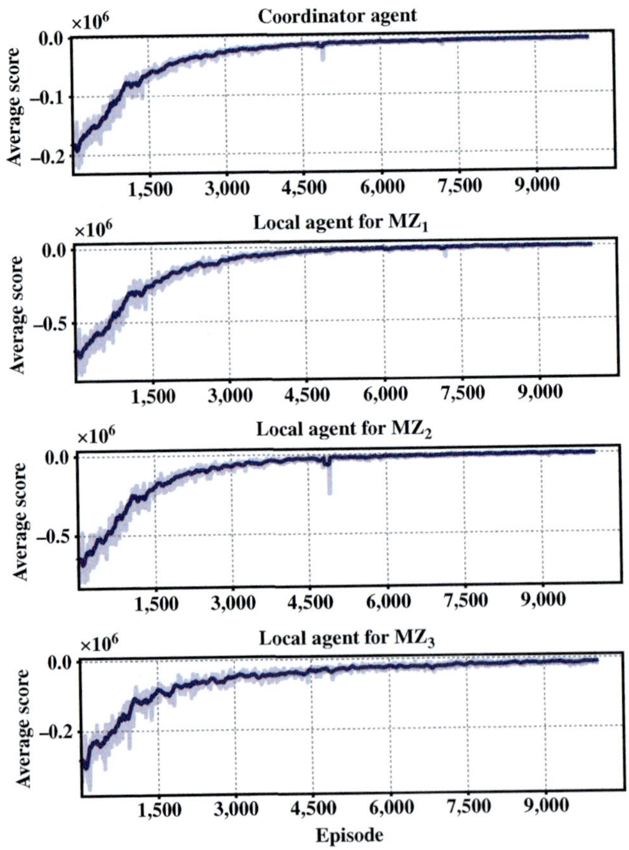


**Figure 9.2** Evolution of the agents' average scores during training of the observer-based SCMARL framework, aggregated over 10,000 episodes and averaged across 10 independent runs.

the agents exhibit steady performance improvement, as evidenced by the rising trend in the average score curves. This upward trend indicates effective policy updates as the agents learn to map belief states to irrigation actions that maximize their respective cumulative rewards.

Beyond the 5,000th episode, the trajectories begin to plateau, suggesting convergence of the learned policies. The flattening of the curves implies that the agents reach a point of policy stability, where further updates result in marginal performance improvements. This behavior reflects successful policy learning under the POMDP framework, with the EKF providing belief-state estimates that are adequately informative for decision-making. The stability of the performance curves also indicates that the EKF-based belief states capture the underlying soil moisture dynamics with adequate accuracy.

## 9.5 Field Implementation of the Integrated Scheduling and Control Layers

Following the training phase of the observer-based SCMARL agents, the framework was integrated with the advanced control layer to enable full-field irrigation scheduling and control for Quadrant 3 of the study area. Specifically, three MPC controllers, one per MZ, were deployed to track the daily irrigation schedules generated by their corresponding local agents within the SCMARL framework.

This integrated implementation was carried out over an entire growing season, under the assumption of uniform wheat cultivation across the quadrant. The field-scale evaluation spanned from May 15, 2022 to September 4, 2022. For the scheduling layer, which operates on a daily basis, weather data relevant to this period are shown in Figure A.3 of Appendix A.3. Hourly temperature and rainfall data were sourced from the Alberta Climate Information Service Data Products & Tools portal (https://www.acis.alberta.ca/). Hourly reference evapotranspiration values were obtained from the methodology described in Singer et al. [119].

Note that although a 30-day scheduling horizon was employed during agent training, the field execution of the trained SCMARL framework was carried out on a daily basis (i.e., with a 1-day horizon). This design choice is supported by two key considerations. First, each agent uses a learned policy that maps the current system state directly to action, without relying on multiday forecasts. Second, in the execution phase, agents are evaluated daily using updated soil moisture estimates and current weather data. Since the agent inputs do not explicitly include long-term forecasts, and due to the fact the control layer operates within a 1-day horizon, there is no added benefit to extending the evaluation period beyond a single day.

At the start of the field implementation, the observer-based SCMARL framework was evaluated using the estimated soil moisture profile and the weather conditions for the first day. If the coordinator agent returned a "no" irrigation decision, the lower-level controller was not activated. In this case, no irrigation was applied, and hourly soil moisture measurements—assumed to represent the average moisture content in the top 25 cm of the soil profile—were assimilated using the EKF to update the moisture estimates for each MZ. The EKF configuration used here followed the design presented in Appendix D.

If the coordinator's decision was "yes," the control layer was engaged. The daily irrigation amounts prescribed by the local agents, along with the corresponding root zone soil moisture levels observed in the training environment, were used to design the objectives of the three MPCs. Each MPC was tuned using weights $\lambda_\theta = 10,000$ and $\lambda_u = 50$. The evaluation of each MPC starts with a 24-hour horizon and uses the current soil moisture estimate as the initial state, along with hourly weather forecasts for the day. Following the first MPC execution, only the computed irrigation input for the first hour was applied. The true soil response was simulated using three independent 1D Richards' equation models to reflect actual field dynamics. To reflect variability in field conditions, Gaussian noise with zero mean and a standard deviation of 0.0007 was incorporated into the simulation of the actual field conditions. After one hour, new soil moisture observations (averaged over the top 25 cm) were collected and used to update the state estimate with the EKF. Sensor noise, which was drawn from a Gaussian distribution with a mean of 0 and a standard deviation of 0.0008, was added to the moisture observations before the EKF step. With the updated state estimate, the MPC was reevaluated using an updated 23-hour forecast, a revised irrigation budget (reflecting the amount already applied), and a shortened control horizon of 23 hours. This cycle of state update, MPC reevaluation, and input application was repeated each hour until the end of the day. This integrated scheduling and control procedure, which consists of daily SCMARL policy evaluation followed by shrinking-horizon MPC execution on irrigation days, was repeated throughout the entire evaluation period.

To evaluate the performance of the integrated scheme, the total amount of irrigation applied by the combined scheduling and control layers was compared to the total amount prescribed by the scheduling layer alone. Additionally, the evolution of root zone soil moisture under the applied irrigation schedule was compared to the target root zone moisture trajectory generated by the scheduler. Finally, the irrigation water use efficiency (IWUE) achieved by the integrated scheme was calculated and compared to that of the scheduler-only approach.

### 9.5.1 Results and Discussion

Figure 9.3 illustrates the evolution of root zone soil moisture in the three MZs under the combined scheduling and control framework. Each plot shows the actual soil moisture trajectory alongside the corresponding reference trajectory prescribed by the scheduling layer. The plots indicate that the lower-level controller is generally effective in tracking the reference trajectories derived from the scheduler.

To quantify tracking performance, Figure 9.4 presents the tracking error, defined as the difference between the actual and reference soil moisture trajectories in each MZ. The maximum observed tracking errors were approximately 0.007, 0.0075, and 0.010 $m^3/m^3$ in MZs 1, 2, and 3, respectively. These relatively small deviations confirm that the lower-level controller successfully refines the irrigation amounts while remaining closely aligned with the scheduler's prescriptions.

Table 9.1 provides a comparative summary of key performance metrics under the combined framework versus the scheduler-only implementation. While both frameworks achieved similar predicted yields (2.52 vs. 2.51 $kg/m^2$), the combined approach resulted in 8.8% less total irrigation. This reduction translated into a 10.1% improvement in IWUE,

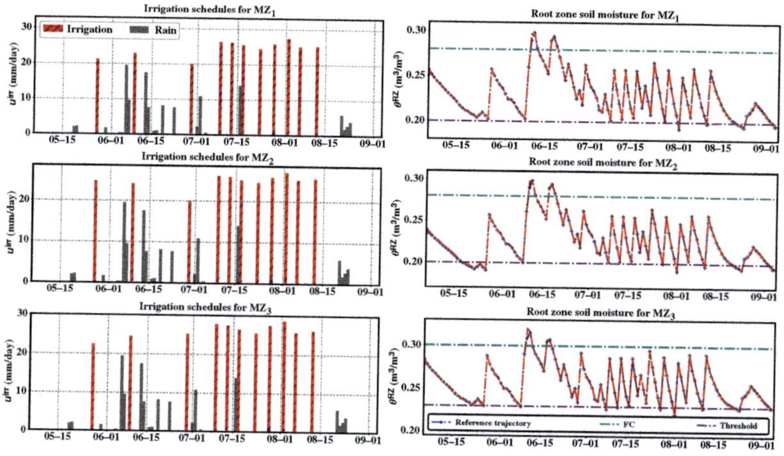


**Figure 9.3** Irrigation prescriptions and root zone soil moisture dynamics resulting from the integrated SCMARL scheduler and MPC-based lower-level controller.

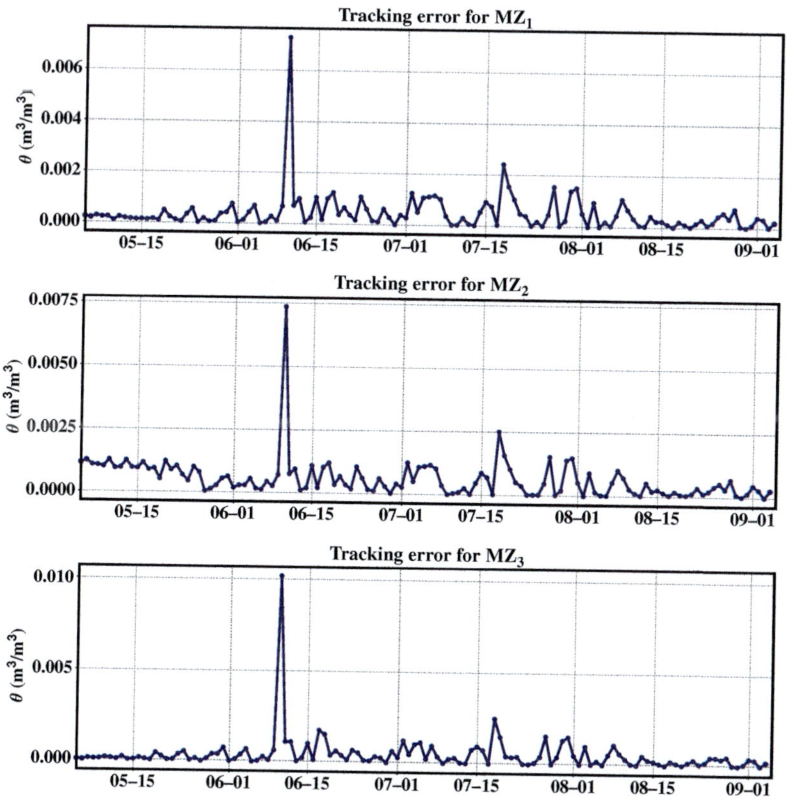


**Figure 9.4** Trajectory of the error between the reference soil moisture trajectory of the scheduler and the root zone soil moisture trajectory resulting from the schedules produced by the integrated scheduler and controller.

**Table 9.1** Comparative analysis of performance metrics for the integrated (scheduler and controller) scheme versus the scheduler-only during field implementation.

| Metric | Scheduler and control | Scheduler only |
|---|---|---|
| Irrigation for $MZ_1$ | 0.27 | 0.29 |
| Irrigation for $MZ_2$ | 0.27 | 0.29 |
| Irrigation for $MZ_3$ | 0.29 | 0.33 |
| Total irrigation (m) | 0.83 [↓ **8.8%**] | 0.91 |
| Predicted yield (kg/m$^2$) | 2.52 [↑ **0.3%**] | 2.51 |
| IWUE (kg/m$^3$) | 3.04 [↑ **10.1%**] | 2.76 |

demonstrating that the incorporation of a high-resolution, feedback-based control layer enhances water efficiency without compromising crop productivity.

Overall, these results suggest that integrating an MPC layer with the SCMARL-based scheduler allows for more precise irrigation adjustments in response to short-term environmental variability, leading to improved resource efficiency in large-scale irrigation systems.

It is also important to consider the computational aspects of the combined scheduling and control framework. As previously discussed, a separate MPC is assigned to each MZ. Since coordination across zones is already handled by the coordinator agent in the scheduling layer, the MPCs for individual MZs can be evaluated independently. This modular structure enables the use of parallel computing techniques to solve the control problems concurrently, significantly improving computational efficiency.

Moreover, the control decisions in the lower layer involve only continuous variables, resulting in a nonlinear program with moderate computational complexity. This allows for the use of physics-based models, such as the 1D Richards' equation, as predictive models within each MPC. In scenarios where computational resources are constrained, surrogate models can be used to replace the Richards' equation.

## 9.6 Conclusion

This chapter introduced a hierarchical framework that integrates a high-level scheduling layer with a low-level control layer to enhance precision in irrigation decision-making for large-scale agricultural systems. The scheduling layer was built on an SCMARL architecture, developed within a partially observable MDP framework. To handle partial observability, an EKF was employed to estimate belief states that represent the soil moisture conditions in each MZ. The learning outcomes demonstrated that the EKF effectively generated belief states that closely approximated true soil moisture distributions, enabling the agents to learn effective policies.

The scheduling layer determined, on a daily basis, whether irrigation should be applied and, if so, prescribed the corresponding amounts for each MZ. These daily prescriptions served as setpoints for the control layer, which was implemented using an MPC strategy. The MPC operated in a shrinking horizon fashion, allowing it to integrate updated hourly weather forecasts and evolving soil moisture estimates. This design ensured that hourly irrigation decisions remained consistent with the daily schedules prescribed by the scheduler.

Field-scale implementation of the integrated framework showed that the combination of scheduling and control layers led to improvements in irrigation efficiency. The system achieved a reduction in total water use and an increase in IWUE while maintaining comparable crop yields relative to a scheduler-only approach. These results highlight the benefits of combining long-term reinforcement learning-based scheduling with short-term model-based control to deliver both resource-efficient and adaptive irrigation decisions.

# 10

# Future Directions

While the methods presented in the preceding chapters show potential, further refinements are necessary to enhance their performance and enable broader deployment. This chapter identifies several research opportunities aimed at advancing these contributions. It begins with possible extensions to the estimation framework for soil moisture and hydraulic parameters discussed in Chapters 4 and 5. It then considers ways to improve the learning-based multi-agent model predictive control (MPC) scheduler introduced in Chapter 7. The subsequent section examines research avenues for scaling the semi-centralized multi-agent reinforcement learning (SCMARL) strategy from Chapter 8. The chapter concludes with prospective enhancements to the hierarchical scheduling and control framework presented in Chapter 9.

## 10.1 Advancing Soil Moisture and Hydraulic Parameter Estimation

A critical next step in advancing the estimation method from Chapter 4 involves expanding its validation across a broader range of environmental conditions, particularly under extreme moisture conditions. The initial analysis did not incorporate such conditions due to limited observational data representing dry or saturated moisture conditions. Addressing this gap will require the design and implementation of targeted experiments that capture soil moisture under these extreme scenarios. Incorporating these data into the estimation method will provide a more comprehensive assessment of the framework's accuracy and generalizability under conditions not previously examined.

Unlike prior approaches [120, 121] that addressed soil hysteresis by applying separate hydraulic parameter sets for wet and dry conditions, the method introduced in Chapter 4 employed a unified set of parameters for all soil moisture states. This was achieved through a recursive estimation process that continuously refined the parameter estimates using real-time soil moisture data, effectively allowing the model to adapt to hysteresis behavior without explicitly modeling distinctions in the soil moisture states. However, future research should include a direct comparison with frameworks that utilize distinct parameter sets for different moisture phases. Such investigations would clarify the relative benefits and limitations of real-time parameter adaptation versus conventional hysteresis-specific modeling strategies.

*Precision Irrigation for Agriculture: Integrating Machine Learning and Optimal Control Strategies*, First Edition. Bernard Twum Agyeman and Jinfeng Liu.

Chapter 4 limited its validation efforts to the estimated saturated hydraulic conductivity due to the lack of laboratory measurements for the other parameters at the study location. To achieve a more complete assessment of the proposed estimation framework, future investigations should pursue validation of additional parameters when reference measurements become available.

Finally, it is worth recalling that the performance-triggered model reduction framework in Chapter 5 was applied for soil moisture estimation. A natural extension of this work would involve leveraging the same framework for the joint estimation of both soil moisture and soil hydraulic parameters.

## 10.2 Advancing the Design of Multi-agent MPC-based Schedulers

The use of parameters obtained through offline estimation significantly contributed to reducing parametric uncertainty and alleviating the mismatch between the model and the actual field. Although this strategy provides a reliable foundation for initial deployment, it lacks the responsiveness necessary to adapt to evolving soil conditions. A flexible alternative would be to implement online estimation of the parameters, which will allow for a continuous refinement of hydraulic parameters using real-time measurements. However, this approach may impose substantial computational demands. An attractive compromise lies in the use of offset-free model predictive control, which augments the system dynamics with a disturbance model. This allows the controller to correct for unmodeled dynamics and external disturbances without explicitly reestimating parameters. For details on the formulation and implementation of offset-free MPC, see the approach described in [122].

Emerging transformer-based models for time-series analysis offer new opportunities for improving prediction in irrigation scheduling. Unlike LSTMs, which inform moisture predictions in the current framework, transformers are well-suited for learning long-range dependencies and learning complex patterns in data. Future work could explore replacing LSTMs with transformer architectures to evaluate gains in robustness and accuracy.

## 10.3 Advancing Semi-centralized Multi-agent RL Applications

Although initially applied to a specific field configuration, the structure of the SCMARL framework is sufficiently flexible to be adapted for other agricultural contexts involving different crops, soil types, and climatic conditions. Such adaptations would require local calibration of the soil model using appropriate hydraulic parameters, as well as adjustments to crop-specific coefficients and environmental inputs. Nonetheless, transferring agent policies trained in one environment directly to another is generally unreliable, as their behavior is tightly coupled to the training domain's soil and weather dynamics. Rather than starting from scratch in each new scenario, future research could investigate transfer learning as a strategy to reuse knowledge across similar tasks. This would help reduce the computational effort and data required when adapting the framework to unfamiliar settings. For an overview of how transfer learning may be used within MARL systems, see the work by [116].

In both the decentralized and semi-centralized MARL frameworks, agent training was conducted without explicitly including weather forecast data in the observation space. Although this choice reduces the complexity of the input, it may constrain the agents' responsiveness to evolving environmental conditions. Incorporating forecasted weather variables into the agent's state representation could allow for more informed and anticipatory decision-making. Future studies could investigate how embedding short- to medium-term forecasts improves performance.

## 10.4 Advancing the Integration of Scheduling and Control

With increasing concerns over water scarcity, seasonal caps on total irrigation water allocations are becoming more common. While the designs presented in this book incorporated daily limits on irrigation amounts, they did not explicitly account for seasonal water constraints. It is important to note that enforcing daily limits does not automatically ensure compliance with seasonal allocation limits, as the relationship between the two is not straightforward. Incorporating seasonal water allocations into the scheduling layer would be beneficial. Doing so would require extending the scheduling horizon to cover the entire growing season and adopting robust optimization strategies capable of handling uncertainties associated with long-term planning.

In the implementation presented in this book, a single MPC was assigned to a representative plot within each management zone (MZ), and the resulting irrigation amounts were uniformly applied to all other plots within that zone. While this constitutes a form of variable rate irrigation (VRI) at the zone level, it does not fully exploit the fine-grained variability present within each MZ. A more complete realization of VRI would involve computing irrigation decisions for all individual plots across the field, based on their unique soil moisture states. The hierarchical framework introduced in this book offers a scalable and computationally tractable approach to enabling such high-resolution control. Specifically, the scheduling layer can continue to focus on representative plots within each MZ, such as the driest plot or one that reflects average moisture conditions, thereby simplifying coordination decisions. Meanwhile, the control layer, implemented as a nonlinear program, can be expanded to include all plots within a given MZ. This formulation allows each MPC to generate spatially differentiated irrigation amounts across plots, based on detailed soil moisture profiles. Moreover, because the control decisions are continuous and handled locally within each MZ, even higher-order versions of the Richards' equation or other detailed soil models can be used without compromising the overall computational budget.

# A

# Relevant Crop and Weather Data

## A.1 Case Studies of Chapters 4 and 5

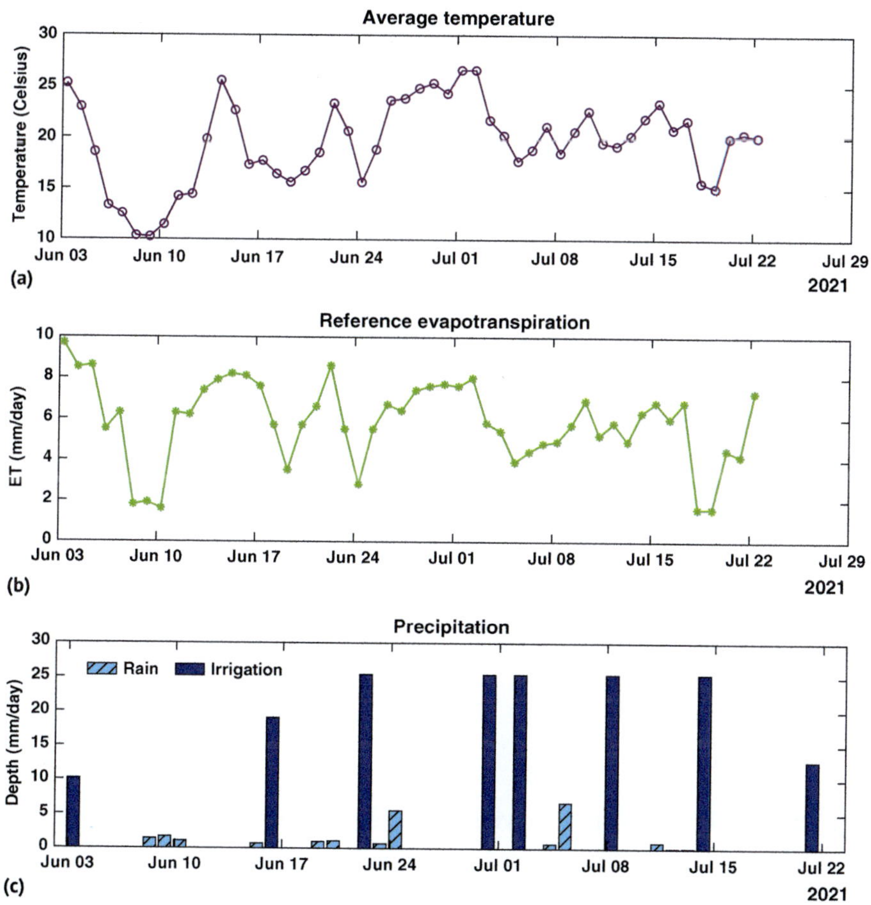


**Figure A.1** Daily trends of average temperature (a), reference evapotranspiration (b), and precipitation (c) over the study period. *Source*: [79]/arXiv/CC BY 4.0.

*Precision Irrigation for Agriculture: Integrating Machine Learning and Optimal Control Strategies*, First Edition. Bernard Twum Agyeman and Jinfeng Liu.

## A.2 Case Studies of Chapter 6

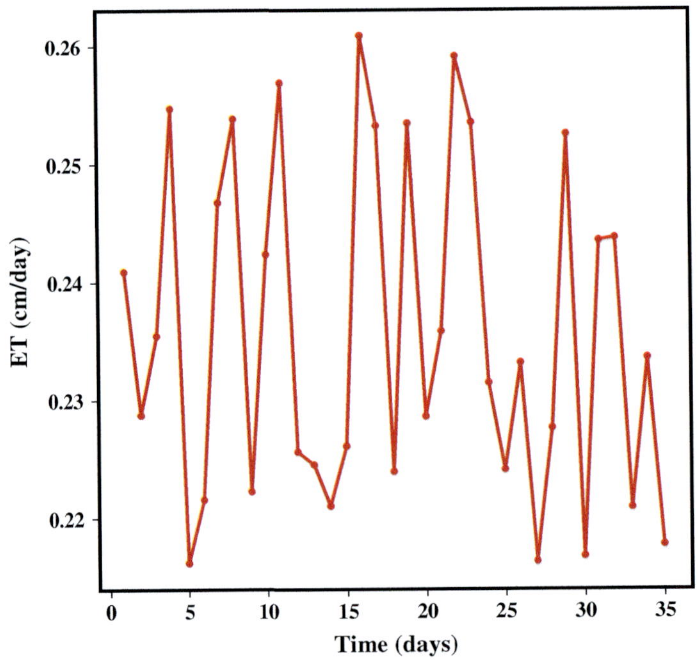


**Figure A.2** Evolution of reference evapotranspiration during the simulation window.

## A.3 Temperature and Evapotranspiration for Case Studies of Chapters 7–9

## A.4 Crop Coefficient Relations

### A.4.1 Barley

The crop coefficient ($K_c$) for barley is computed using the following relation [123]:

$$K_c(g) = 0.04217 + 0.001508g + (4.89 \times 10^{-6})\,g^2 - (8.69 \times 10^{-9})\,g^3 + (2.49 \times 10^{-12})\,g^4 \tag{A.1}$$

The term g refers to the cumulative growing-degree days (GDD) and it is calculated as follows:

$$\text{GDD} = T_{\text{avg}} - T_{\text{base}} \tag{A.2}$$

where $T_{\text{avg}}$ represents the daily mean temperature. The base temperature, which refers to the temperature (5 °C) below which crop growth stops, is denoted as $T_{\text{base}}$.

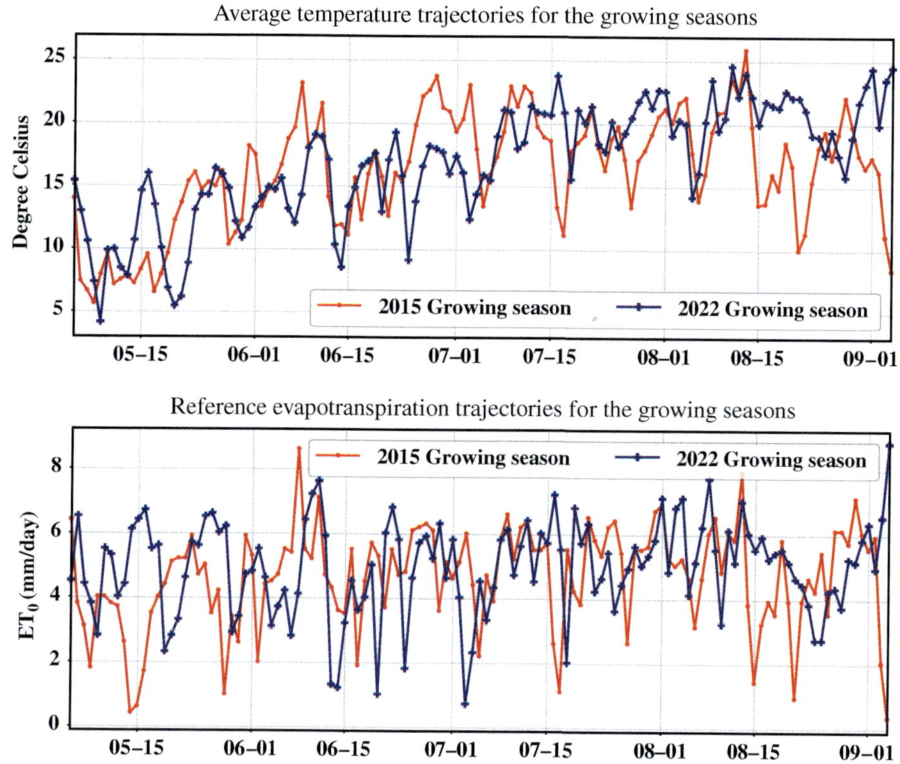


**Figure A.3** Daily profiles of reference evapotranspiration and mean temperature.

## A.4.2 Spring Soft Wheat

The $K_c$ of soft spring wheat is computed using the following relation [123]:

$$K_c(g) = -0.0207 + 0.00266g + \left(4.7 \times 10^{-8}\right) g^2 - \left(2.0 \times 10^{-9}\right) g^3 + \left(2.70 \times 10^{-13}\right) g^4 \tag{A.3}$$

where $g$ denotes the cumulative GDD. The GDD is the difference between the daily average temperature $T_{avg}$ and the base temperature $T_{base}$, which is taken as (5 °C).

# B

# Relevant Soil Hydraulic Parameters

## B.1 Nominal Parameter Maps Used in Sensitivity Analysis

A field-scale soil texture analysis was performed at the Lethbridge Polytechnic Research Farm in Alberta, Canada, to support the derivation of nominal hydraulic parameters. Samples were collected from 60 points at a uniform depth of 30 cm, and laboratory tests were conducted to determine the proportions of sand, silt, and clay. Using these texture measurements, hydraulic parameters were estimated through established pedotransfer models described in [86] and [124]. To generate continuous spatial maps of these parameters, kriging interpolation was applied across the field domain.

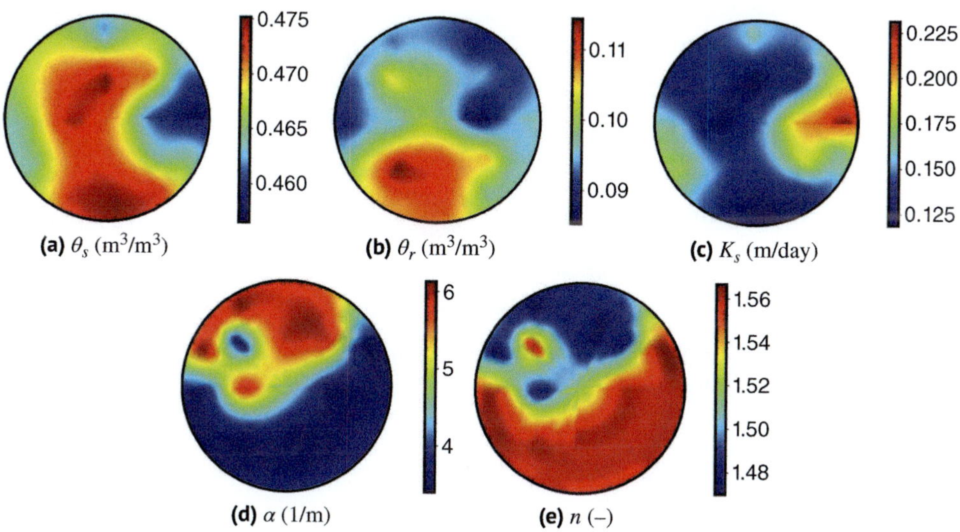


**Figure B.1** Nominal values of soil hydraulic parameters across the field.
*Source*: [79]/arXiv/CC BY 4.0.

*Precision Irrigation for Agriculture: Integrating Machine Learning and Optimal Control Strategies*,
First Edition. Bernard Twum Agyeman and Jinfeng Liu.

## B.2 Soil Parameters Used in Validation of Chapter 4

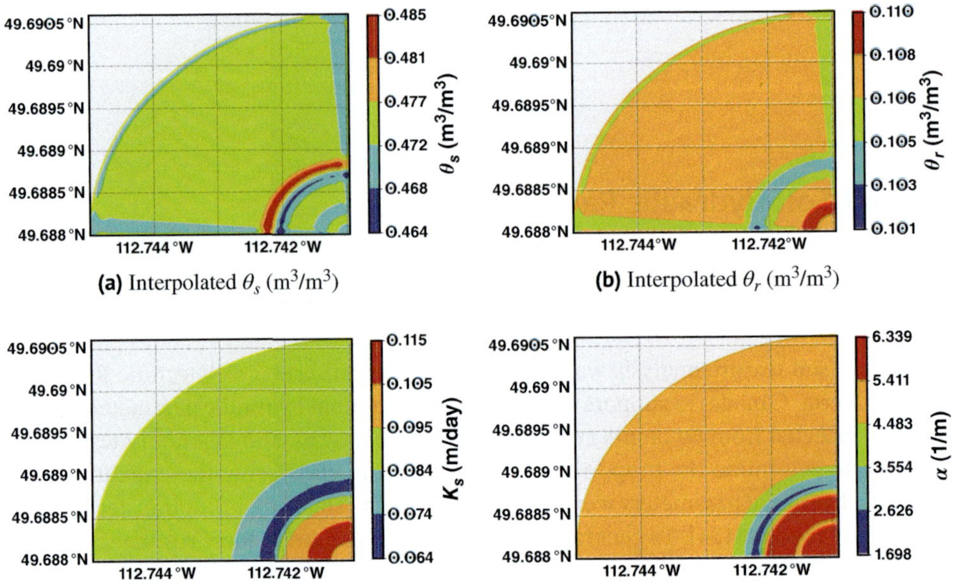


**(a)** Interpolated $\theta_s$ (m$^3$/m$^3$)

**(b)** Interpolated $\theta_r$ (m$^3$/m$^3$)

**(c)** Interpolated $K_s$ (m/day)

**(d)** Interpolated $\alpha$ (1/m)

**(e)** Interpolated $n$ (–)

**Figure B.2** Field-scale interpolation of soil hydraulic parameters based on texture survey data. *Source*: [79]/arXiv/CC BY 4.0.

# C

# Crop Yield Calculations

Predictions of crop yield are determined as follows [123]:

$$Y_a = Y_m \left[1 - k_y + \left(k_y \times \frac{\mathrm{ET}_c}{\mathrm{ET}_m}\right)\right] \tag{C.1}$$

The term $Y_a$ denotes the actual crop yield (kg/m$^2$), $Y_m$ is the maximum achievable yield under ideal conditions (kg/m$^2$), and $k_y$ is a dimensionless yield response coefficient specific to the crop. For spring soft wheat, $Y_m$ and $k_y$ were set to 0.88 kg/m$^2$ and 1.15, respectively—values calibrated for the study area as reported in [123]. The variable $\mathrm{ET}_c$ represents the actual seasonal crop evapotranspiration (mm), while $\mathrm{ET}_m$ is the maximum seasonal evapotranspiration under non-limiting water conditions (mm). This formulation assumes that water availability is the primary limiting factor for yield. The relationship between $\mathrm{ET}_c$ and $\mathrm{ET}_m$ follows the expression given in [66]:

$$\mathrm{ET}_c = \mathcal{K}(\theta_\mathrm{v})\mathrm{ET}_m \tag{C.2}$$

where $\mathcal{K}(\cdot)$ denotes the water stress coefficient, which depends on the volumetric soil moisture content $\theta_v$. The function $\mathcal{K}(\cdot)$ is given by:

$$\mathcal{K}(\theta_\mathrm{v}) = \begin{cases} 0 & \theta_{\mathrm{v}_1} \leq \theta_\mathrm{v} \\ \dfrac{\theta_\mathrm{v} - \theta_{\mathrm{v}_1}}{\theta_{\mathrm{v}_2} - \theta_{\mathrm{v}_1}} & \theta_{\mathrm{v}_1} \leq \theta_\mathrm{v} \leq \theta_{\mathrm{v}_2} \\ 1 & \theta_{\mathrm{v}_2} \leq \theta_\mathrm{v} \leq \theta_{\mathrm{v}_3} \\ \dfrac{\theta_\mathrm{v} - \theta_{\mathrm{v}_w}}{\theta_{\mathrm{v}_2} - \theta_{\mathrm{v}_w}} & \theta_{\mathrm{v}_w} \leq \theta_\mathrm{v} \leq \theta_{\mathrm{v}_2} \end{cases} \tag{C.3}$$

where $\theta_{\mathrm{v}_1}$ represents the volumetric water content at the anaerobic threshold, and $\theta_{\mathrm{v}_2}$ and $\theta_{\mathrm{v}_3}$ define the interval within which plant water uptake is optimal. Specifically, $\theta_{\mathrm{v}_3} = \bar{\nu}$ (the upper bound of the target soil moisture range) and $\theta_{\mathrm{v}_2} = \underline{\nu}$ (the lower bound). The term $\theta_{\mathrm{v}_w}$ corresponds to the permanent wilting point, denoted as $\theta_{\mathrm{wp}}$. The anaerobic point $\theta_{\mathrm{v}_1}$ was determined as the water content associated with a pressure head of $-0.1$ m, following the approach in [20].

*Precision Irrigation for Agriculture: Integrating Machine Learning and Optimal Control Strategies*,
First Edition. Bernard Twum Agyeman and Jinfeng Liu.

# D

# Extended Kalman Filter Design

The estimation process begins with an initial guess of the state vector, denoted by $\hat{x}_0^+$. The associated error covariance matrix is set as $P_0^+ = 15.9 \times \mathbb{I}_{21}$, where $\mathbb{I}_{21}$ is the $21 \times 21$ identity matrix. The covariance matrix for the process noise is defined as $Q = 0.05 \times \mathbb{I}_{21}$, while the measurement noise covariance is assigned a fixed value of $R = 19.25$.

At time step $k+1$, $\hat{x}_k^+$ and $P_k^+$ are propagated through the model as follows:

$$\hat{x}_{k+1}^- = \mathcal{F}_d(\hat{x}_k^+, u_k, \boldsymbol{\phi}) \tag{D.1}$$

$$P_{k+1}^- = A_k P_k^+ A_k^T + Q \tag{D.2}$$

where $A_k = \frac{\partial \mathcal{F}_d}{\partial x}\big|_{\hat{x}_k^+, u_k}$

With the measurement $o_{k+1}$ at time step $k+1$, $\hat{x}_{k+1}^-$ and $P_{k+1}^-$ are updated as follows:

$$\hat{y}_{k+1}^- = \mathcal{H}(\hat{x}_{k+1}^-, \boldsymbol{\phi}) \tag{D.3}$$

$$G_{k+1} = P_{k+1}^- C_{k+1}^T [C P_{k+1}^- C_{k+1}^T + R]^{-1} \tag{D.4}$$

$$\hat{x}_{k+1}^+ = \hat{x}_{k+1}^- + G_{k+1}[o_{k+1} - \mathcal{M}\hat{y}_{k+1}^-] \tag{D.5}$$

$$P_{k+1}^+ = [I - G_{k+1} C_{k+1}] P_{k+1}^- \tag{D.6}$$

$$\hat{y}_{k+1}^+ = \mathcal{H}(\hat{x}_{k+1}^+, \boldsymbol{\phi}) \tag{D.7}$$

where $\mathcal{M}$ is a selection matrix used to extract the relevant components of the predicted output vector $\hat{y}$ that contribute to the observed soil moisture measurement $o$ and $C_{k+1} = \frac{\partial \mathcal{H}}{\partial x}\big|_{\hat{x}_{k+1}^+}$.

*Precision Irrigation for Agriculture: Integrating Machine Learning and Optimal Control Strategies*, First Edition. Bernard Twum Agyeman and Jinfeng Liu.

# E

# Relevant Parameters and Simulation Setup

## E.1 Richards' Equation

**Table E.1** Details of the simulation setup used for solving the 1D Richards' equation.

| | |
|---|---|
| **Total depth** | 1.0 m |
| **Time step size** | 1,800 s |
| **Temporal discretization** | Backward differentiation formula |
| **Number of spatial nodes** | 31 |
| **Spatial discretization** | Top 0.50 m with 21 nodes. Bottom 0.50 m with 11 nodes |
| **Process uncertainty** | Normal distribution with mean of 0 and standard deviation of 5e-04 |

## E.2 LSTM Network

**Table E.2** Architecture and configuration of the LSTM network.

| | |
|---|---|
| **Loss** | **Mean squared error** |
| **Optimizer** | Adam |
| **Number of epochs** | 40 |
| **Learning rate** | 0.0001 |
| **Number of hidden layers** | 2 |
| **Sequence length** | 5 |
| **Number of LSTM units in a hidden layer** | 400 |

The values listed above were obtained through empirical experimentation. However, it is likely that a more systematic hyperparameter tuning process could yield improved performance outcomes.

*Precision Irrigation for Agriculture: Integrating Machine Learning and Optimal Control Strategies*, First Edition. Bernard Twum Agyeman and Jinfeng Liu.

## E.3 Hybrid PPO Agent

**Table E.3** Specification of hyperparameters for the PPO.

| Hyperparameter | Value |
|---|---|
| Discount factor | 0.99 |
| Clipping parameter | 0.25 |
| Learning rate | 0.0001 |
| Minibatch size | 64 |
| Number of episodes | 300,000 |
| Number of epochs | 10 |
| Generalized advantage estimation parameter | 0.97 |
| Entropy coefficient | 0.01 |
| Horizon | 30 |

The values listed above were obtained through empirical experimentation. It is likely that a more systematic hyperparameter tuning process could yield improved performance outcomes.

## E.4 Parameters and Weights

**Table E.4** Values assigned to the weights and parameters of formulations $\mathbb{P}_{\mathrm{N}}(y, d)$ and $\mathbb{P}_{\mathrm{M,N}}(\boldsymbol{Y}, d)$.

| Weight/Parameter | Value |
|---|---|
| $\underline{Q}$ | 2.0e8 |
| $\bar{Q}$ | 2.2e8 |
| $R_c$ | 1.0e3 |
| $R_u$ | 9.0e3 |
| $\theta_{\mathrm{fc}}/\bar{\nu}$ ($\mathrm{MZ}_1$, $\mathrm{MZ}_2$, $\mathrm{MZ}_3$) | 0.280, 0.280, 0.300 |
| $\underline{\nu}$ ($\mathrm{MZ}_1$, $\mathrm{MZ}_2$, $\mathrm{MZ}_3$) for MAD = 65% | 0.176, 0.176, 0.209 |
| $\underline{\nu}$ ($\mathrm{MZ}_1$, $\mathrm{MZ}_2$, $\mathrm{MZ}_3$) for MAD = 40% | 0.216, 0.216, 0.244 |
| $\theta_{\mathrm{pwp}}$ ($\mathrm{MZ}_1$, $\mathrm{MZ}_2$, $\mathrm{MZ}_3$) | 0.120, 0.120, 0.160 |

The cost parameters were treated as adjustable variables. To determine suitable values, multiple configurations were simulated, with each involving different cost parameter settings. Performance was assessed under these variations to identify configurations that yielded the most favorable trade-offs between water use and scheduling objectives. The moisture thresholds $\bar{\nu}$ and $\underline{\nu}$ were computed based on field capacity ($\theta_{\text{fc}}$) and wilting point ($\theta_{\text{wp}}$) values. These values were obtained from [106].

## References


1 UN Water, WWAP (United Nations World Water Assessment Programme), *Facing the Challenges, Case Studies and Indicators*. UNESCO, Paris, 2015.

2 S. L. Shah, B. R. Bakshi, J. Liu, C. Georgakis, B. Chachuat, R. D. Braatz, and B. R. Young, "Meeting the challenge of water sustainability: The role of process systems engineering," *AIChE Journal*, vol. 67, no. 2, p. e17113, 2021.

3 K. C. DeJonge, S. Taghvaeian, T. J. Trout, and L. H. Comas, "Comparison of canopy temperature-based water stress indices for maize," *Agricultural Water Management*, vol. 156, pp. 51–62, 2015.

4 C. Padilla-Díaz, C. Rodriguez-Dominguez, V. Hernandez-Santana, A. Perez-Martin, and J. Fernández, "Scheduling regulated deficit irrigation in a hedgerow olive orchard from leaf turgor pressure related measurements," *Agricultural Water Management*, vol. 164, pp. 28–37, 2016.

5 Y. Osroosh, R. T. Peters, C. S. Campbell, and Q. Zhang, "Automatic irrigation scheduling of apple trees using theoretical crop water stress index with an innovative dynamic threshold," *Computers and Electronics in Agriculture*, vol. 118, pp. 193–203, 2015.

6 A. Bartlett, A. A. Andales, M. Arabi, and T. Bauder, "A smartphone app to extend use of a cloud-based irrigation scheduling tool," *Computers and Electronics in Agriculture*, vol. 111, pp. 127–130, 2015.

7 C. Hedley and I. Yule, "A method for spatial prediction of daily soil water status for precise irrigation scheduling," *Agricultural Water Management*, vol. 96, no. 12, pp. 1737–1745, 2009.

8 R. Thompson, M. Gallardo, L. Valdez, and M. Fernández, "Using plant water status to define threshold values for irrigation management of vegetable crops using soil moisture sensors," *Agricultural Water Management*, vol. 88, no. 1–3, pp. 147–158, 2007.

9 J. Gutiérrez, J. F. Villa-Medina, A. Nieto-Garibay, and M. Á. Porta-Gándara, "Automated irrigation system using a wireless sensor network and GPRS module," *IEEE Transactions on Instrumentation and Measurement*, vol. 63, no. 1, pp. 166–176, 2013.

10 J. Haule and K. Michael, "Deployment of wireless sensor networks (WSN) in automated irrigation management and scheduling systems: A review," in *Proceedings of the 2nd Pan African International Conference on Science, Computing and Telecommunications (PACT 2014)*, pp. 86–91, IEEE, 2014.



*Precision Irrigation for Agriculture: Integrating Machine Learning and Optimal Control Strategies*, First Edition. Bernard Twum Agyeman and Jinfeng Liu.

**11** L. N. Sethi, S. N. Panda, and M. K. Nayak, "Optimal crop planning and water resources allocation in a coastal groundwater basin, Orissa, India," *Agricultural Water Management*, vol. 83, no. 3, pp. 209–220, 2006.

**12** D. Khare, M. Jat, and J. D. Sunder, "Assessment of water resources allocation options: Conjunctive use planning in a link canal command," *Resources, Conservation and Recycling*, vol. 51, no. 2, pp. 487–506, 2007.

**13** R. G. Perea, I. F. García, M. M. Arroyo, J. R. Díaz, E. C. Poyato, and P. Montesinos, "Multiplatform application for precision irrigation scheduling in strawberries," *Agricultural Water Management*, vol. 183, pp. 194–201, 2017.

**14** S. O. Lopes, F. A. Fontes, R. M. Pereira, M. de Pinho, and A. M. Gonçalves, "Optimal control applied to an irrigation planning problem," *Mathematical Problems in Engineering*, vol. 2016, no. 1, p. 5076879, 2016.

**15** D. C. H. Nguyen, J. C. Ascough II, H. R. Maier, G. C. Dandy, and A. A. Andales, "Optimization of irrigation scheduling using ant colony algorithms and an advanced cropping system model," *Environmental Modelling and Software*, vol. 97, pp. 32–45, 2017.

**16** Y. Park, J. S. Shamma, and T. C. Harmon, "A receding horizon control algorithm for adaptive management of soil moisture and chemical levels during irrigation," *Environmental Modelling and Software*, vol. 24, no. 9, pp. 1112–1121, 2009.

**17** D. Delgoda, H. Malano, S. K. Saleem, and M. N. Halgamuge, "Irrigation control based on model predictive control (MPC): Formulation of theory and validation using weather forecast data and AQUACROP model," *Environmental Modelling and Software*, vol. 78, pp. 40–53, 2016.

**18** C. Lozoya, C. Mendoza, L. Mejía, J. Quintana, G. Mendoza, M. Bustillos, O. Arras, and L. Solís, "Model predictive control for closed-loop irrigation," *IFAC Proceedings Volumes*, vol. 47, no. 3, pp. 4429–4434, 2014.

**19** Z. Gu, T. Zhu, X. Jiao, J. Xu, and Z. Qi, "Neural network soil moisture model for irrigation scheduling," *Computers and Electronics in Agriculture*, vol. 180, p. 105801, 2021.

**20** F. Capraro, D. Patino, S. Tosetti, and C. Schugurensky, "Neural network-based irrigation control for precision agriculture," in *2008 IEEE International Conference on Networking, Sensing and Control*, pp. 357–362, IEEE, 2008.

**21** S. Tsang and C. Y. Jim, "Applying artificial intelligence modeling to optimize green roof irrigation," *Energy and Buildings*, vol. 127, pp. 360–369, 2016.

**22** O. Adeyemi, I. Grove, S. Peets, Y. Domun, and T. Norton, "Dynamic neural network modelling of soil moisture content for predictive irrigation scheduling," *Sensors*, vol. 18, no. 10, p. 3408, 2018.

**23** Z. Gu, Z. Qi, R. Burghate, S. Yuan, X. Jiao, and J. Xu, "Irrigation scheduling approaches and applications: A review," *Journal of Irrigation and Drainage Engineering*, vol. 146, no. 6, p. 04020007, 2020.

**24** H. Zhao, L. Di, E. Yu, L. Guo, L. Li, C. Zhang, and H. Li, "A review of scientific irrigation scheduling methods," in *2023 11th International Conference on Agro-Geoinformatics (Agro-Geoinformatics)*, pp. 1–4, IEEE, 2023.

**25** J. Nahar, S. Liu, Y. Mao, J. Liu, and S. L. Shah, "Closed-loop scheduling and control for precision irrigation," *Industrial and Engineering Chemistry Research*, vol. 58, no. 26, pp. 11485–11497, 2019.

**26** A. C. McCarthy, N. H. Hancock, and S. R. Raine, "Simulation of irrigation control strategies for cotton using model predictive control within the VARIwise simulation framework," *Computers and Electronics in Agriculture*, vol. 101, pp. 135–147, 2014.

**27** C. A. Floudas and X. Lin, "Mixed integer linear programming in process scheduling: Modeling, algorithms, and applications," *Annals of Operations Research*, vol. 139, pp. 131–162, 2005.

**28** E. A. Abioye, O. Hensel, T. J. Esau, O. Elijah, M. S. Z. Abidin, A. S. Ayobami, O. Yerima, and A. Nasirahmadi, "Precision irrigation management using machine learning and digital farming solutions," *AgriEngineering*, vol. 4, no. 1, pp. 70–103, 2022.

**29** T. A. Shaikh, T. Rasool, and F. R. Lone, "Towards leveraging the role of machine learning and artificial intelligence in precision agriculture and smart farming," *Computers and Electronics in Agriculture*, vol. 198, p. 107119, 2022.

**30** A. Haghverdi, B. G. Leib, R. A. Washington-Allen, P. D. Ayers, and M. J. Buschermohle, "Perspectives on delineating management zones for variable rate irrigation," *Computers and Electronics in Agriculture*, vol. 117, pp. 154–167, 2015.

**31** M. Monem and S. Hashemi, "Spatial clustering of irrigation networks using *K*-means method (Case study of Ghazvin irrigation network)," *Iran-Water Resources Research*, vol. 7, no. 1, pp. 38–46, 2010.

**32** M. S. Metwally, S. M. Shaddad, M. Liu, R.-J. Yao, A. I. Abdo, P. Li, J. Jiao, and X. Chen, "Soil properties spatial variability and delineation of site-specific management zones based on soil fertility using fuzzy clustering in a hilly field in Jianyang, Sichuan, China," *Sustainability*, vol. 11, no. 24, p. 7084, 2019.

**33** A. Tagarakis, V. Liakos, S. Fountas, S. Koundouras, and T. Gemtos, "Management zones delineation using fuzzy clustering techniques in grapevines," *Precision Agriculture*, vol. 14, pp. 18–39, 2013.

**34** H. Medina, N. Romano, and G. Chirico, "Kalman filters for assimilating near-surface observations into the Richards equation – part 2: A dual filter approach for simultaneous retrieval of states and parameters," *Hydrology and Earth System Sciences*, vol. 18, no. 7, pp. 2521–2541, 2014.

**35** H. Lü, Z. Yu, Y. Zhu, S. Drake, Z. Hao, and E. A. Sudicky, "Dual state-parameter estimation of root zone soil moisture by optimal parameter estimation and extended Kalman filter data assimilation," *Advances in Water Resources*, vol. 34, no. 3, pp. 395–406, 2011.

**36** C. Li and L. Ren, "Estimation of unsaturated soil hydraulic parameters using the ensemble Kalman filter," *Vadose Zone Journal*, vol. 10, no. 4, pp. 1205–1227, 2011.

**37** H. Moradkhani, S. Sorooshian, H. V. Gupta, and P. R. Houser, "Dual state–parameter estimation of hydrological models using ensemble Kalman filter," *Advances in Water Resources*, vol. 28, no. 2, pp. 135–147, 2005.

**38** W. Chen, C. Huang, H. Shen, and X. Li, "Comparison of ensemble-based state and parameter estimation methods for soil moisture data assimilation," *Advances in Water Resources*, vol. 86, pp. 425–438, 2015.

**39** N. Brandhorst, D. Erdal, and I. Neuweiler, "Soil moisture prediction with the ensemble Kalman filter: Handling uncertainty of soil hydraulic parameters," *Advances in Water Resources*, vol. 110, pp. 360–370, 2017.

**40** H. Zhang, H.-J. Hendricks Franssen, X. Han, J. A. Vrugt, and H. Vereecken, "State and parameter estimation of two land surface models using the ensemble Kalman filter and the particle filter," *Hydrology and Earth System Sciences*, vol. 21, no. 9, pp. 4927–4958, 2017.

**41** S. Bo and J. Liu, "A decentralized framework for parameter and state estimation of infiltration processes," *Mathematics*, vol. 8, no. 5, p. 681, 2020.

**42** S. Bo, S. R. Sahoo, X. Yin, J. Liu, and S. L. Shah, "Parameter and state estimation of one-dimensional infiltration processes: A simultaneous approach," *Mathematics*, vol. 8, no. 1, p. 134, 2020.

**43** M. P. Saccomani, S. Audoly, and L. D'Angiò, "Parameter identifiability of nonlinear systems: The role of initial conditions," *Automatica*, vol. 39, no. 4, pp. 619–632, 2003.

**44** K. Z. Yao, B. M. Shaw, B. Kou, K. B. McAuley, and D. Bacon, "Modeling ethylene/butene copolymerization with multi-site catalysts: Parameter estimability and experimental design," *Polymer Reaction Engineering*, vol. 11, no. 3, pp. 563–588, 2003.

**45** T. Jain, P. Garg, P. K. Tiwari, V. K. Kuncham, M. Sharma, and V. K. Verma, "Performance prediction for crop irrigation using different machine learning approaches," in *Examining the Impact of Deep Learning and IoT on Multi-Industry Applications*, pp. 61–79, IGI Global, 2021.

**46** Y. Suzuki, H. Ibayashi, and H. Mineno, "An SVM based irrigation control system for home gardening," in *2013 IEEE 2nd Global Conference on Consumer Electronics (GCCE)*, pp. 365–366, IEEE, 2013.

**47** A. Glória, J. Cardoso, and P. Sebastião, "Sustainable irrigation system for farming supported by machine learning and real-time sensor data," *Sensors*, vol. 21, no. 9, p. 3079, 2021.

**48** H. Shalini and C. Aravinda, "An IoT-based predictive analytics for estimation of rainfall for irrigation," in *International Conference on Artificial Intelligence and Data Engineering*, pp. 1399–1413, Springer, 2019.

**49** H. Niu, T. Zhao, D. Wang, and Y. Chen, "A UAV resolution and waveband aware path planning for onion irrigation treatments inference," in *2019 International Conference on Unmanned Aircraft Systems (ICUAS)*, pp. 808–812, IEEE, 2019.

**50** J. Cardoso, A. Gloria, and P. Sebastiao, "Improve irrigation timing decision for agriculture using real time data and machine learning," in *2020 International Conference on Data Analytics for Business and Industry: Way Towards a Sustainable Economy (ICDABI)*, pp. 1–5, IEEE, 2020.

**51** K. Bhanu, H. Mahadevaswamy, and H. Jasmine, "IoT based smart system for enhanced irrigation in agriculture," in *2020 International Conference on Electronics and Sustainable Communication Systems (ICESC)*, pp. 760–765, IEEE, 2020.

**52** S. Kumar, S. Mishra, and P. Khanna, "Precision sugarcane monitoring using SVM classifier," *Procedia Computer Science*, vol. 122, pp. 881–887, 2017.

**53** H. Han, C. Choi, J. Kim, R. R. Morrison, J. Jung, and H. S. Kim, "Multiple-depth soil moisture estimates using artificial neural network and long short-term memory models," *Water*, vol. 13, no. 18, p. 2584, 2021.

**54** L. Sun, Y. Yang, J. Hu, D. Porter, T. Marek, and C. Hillyer, "Reinforcement learning control for water-efficient agricultural irrigation," in *2017 IEEE International Symposium on Parallel and Distributed Processing with Applications and 2017 IEEE International Conference on Ubiquitous Computing and Communications (ISPA/IUCC)*, pp. 1334–1341, IEEE, 2017.

55 M. Chen, Y. Cui, X. Wang, H. Xie, F. Liu, T. Luo, S. Zheng, and Y. Luo, "A reinforcement learning approach to irrigation decision-making for rice using weather forecasts," *Agricultural Water Management*, vol. 250, p. 106838, 2021.

56 J.-E. Bergez, M. Eigenraam, and F. Garcia, "Comparison between dynamic programming and reinforcement learning: A case study on maize irrigation management," in *Proceedings of the 3rd European Conference on Information Technology in Agriculture (EFITA01), Montpellier (FR)*, pp. 343–348, 2001.

57 Y. Yang, J. Hu, D. Porter, T. Marek, K. Heflin, and H. Kong, "Deep reinforcement learning-based irrigation scheduling," *Transactions of the ASABE*, vol. 63, no. 3, pp. 549–556, 2020.

58 Z. Fan, R. Su, W. Zhang, and Y. Yu, "Hybrid actor-critic reinforcement learning in parameterized action space," *arXiv preprint arXiv:1903.01344*, 2019.

59 E. Wei, D. Wicke, and S. Luke, "Hierarchical approaches for reinforcement learning in parameterized action space," in *AAAI Spring Symposia*, 2018.

60 A.-F. Jimenez, P.-F. Cardenas, A. Canales, F. Jimenez, and A. Portacio, "A survey on intelligent agents and multi-agents for irrigation scheduling," *Computers and Electronics in Agriculture*, vol. 176, p. 105474, 2020.

61 S. Belaqziz, "An agent-based modeling approach for decision-making in gravity irrigation systems," in *2011 International Conference for Internet Technology and Secured Transactions*, pp. 673–680, IEEE, 2011.

62 T. Wanyama and B. Far, "Multi-agent system for irrigation using fuzzy logic algorithm and open platform communication data access," *International Journal of Computer and Information Engineering*, vol. 11, no. 6, pp. 690–695, 2017.

63 D. B. Stephens, *Vadose Zone Hydrology*. CRC Press, Boca Raton, FL, USA, 2018.

64 G. Bonan, *Climate Change and Terrestrial Ecosystem Modeling*. Cambridge University Press, Cambridge, UK, 2019.

65 M. T. Van Genuchten, "A closed-form equation for predicting the hydraulic conductivity of unsaturated soils," *Soil Science Society of America Journal*, vol. 44, no. 5, pp. 892–898, 1980.

66 R. A. Feddes, "Simulation of field water use and crop yield," in *Simulation of Plant Growth and Crop Production*, pp. 194–209, Pudoc, 1982.

67 M. Gupta, P. Srivastava, and T. Islam, "Integrative use of near-surface satellite soil moisture and precipitation for estimation of improved irrigation scheduling parameters," in *Satellite Soil Moisture Retrieval*, pp. 271–288, Elsevier, 2016.

68 S. Al-Khafaf, P. Wierenga, and B. Williams, "Evaporative flux from irrigated cotton as related to leaf area index, soil water, and evaporative demand 1," *Agronomy Journal*, vol. 70, no. 6, pp. 912–917, 1978.

69 J. B. Rawlings, D. Q. Mayne, and M. Diehl, *Model Predictive Control: Theory, Computation, and Design*, vol. 2. Nob Hill Publishing, Madison, WI, USA, 2017.

70 R. S. Sutton and A. G. Barto, *Reinforcement Learning: An Introduction*. MIT Press, Cambridge, MA, USA, 2018.

71 Y. Bengio, I. Goodfellow, and A. Courville. *Deep Learning*, (vol. 1), MIT Press, Cambridge, MA, USA, 2017.

72 S. Hochreiter and J. Schmidhuber, "Long short-term memory," *Neural Computation*, vol. 9, no. 8, pp. 1735–1780, 1997.

73 J. Schulman, F. Wolski, P. Dhariwal, A. Radford, and O. Klimov, "Proximal policy optimization algorithms," *arXiv preprint arXiv:1707.06347*, 2017.

**74** A. Wächter and L. T. Biegler, "On the implementation of primal-dual interior point filter line search algorithm for large-scale nonlinear programming," *Mathematical Programming*, vol. 106, pp. 25–57, 2006.

**75** Gurobi Optimization, LLC, "Gurobi Optimizer Reference Manual," 2024.

**76** I. I. Cplex, "V12. 1: User's manual for CPLEX," *International Business Machines Corporation*, vol. 46, no. 53, p. 157, 2009.

**77** P. Bonami, L. T. Biegler, A. R. Conn, G. Cornuéjols, I. E. Grossmann, C. D. Laird, J. Lee, A. Lodi, F. Margot, and N. Sawaya, "An algorithmic framework for convex mixed integer nonlinear programs," *Discrete Optimization*, vol. 5, no. 2, pp. 186–204, 2008.

**78** E. Orouskhani, B. T. Agyeman, and J. Liu, "Simultaneous estimation of soil moisture and hydraulic parameters for precision agriculture. part A: Methodology," in *2022 IEEE International Symposium on Advanced Control of Industrial Processes (AdCONIP)*, pp. 12–17, IEEE, 2022.

**79** B. T. Agyeman, E. Orouskhani, M. Naouri, W. Appels, M. Wolleben, J. Liu, and S. L. Shah, "Sensitivity-informed parameter selection for improved soil moisture estimation from remote sensing data," *arXiv preprint arXiv:2305.15549*, 2025.

**80** B. T. Agyeman, E. Orouskhani, and J. Liu, "Simultaneous estimation of soil moisture and hydraulic parameters for precision agriculture. part B: Application to a real field," in *2022 IEEE International Symposium on Advanced Control of Industrial Processes (AdCONIP)*, pp. 18–23, IEEE, 2022.

**81** B. T. Agyeman, S. Bo, S. R. Sahoo, X. Yin, J. Liu, and S. L. Shah, "Soil moisture map construction by sequential data assimilation using an extended Kalman filter," *Journal of Hydrology*, vol. 598, p. 126425, 2021.

**82** J. Liu, A. Gnanasekar, Y. Zhang, S. Bo, J. Liu, J. Hu, and T. Zou, "Simultaneous state and parameter estimation: The role of sensitivity analysis," *Industrial & Engineering Chemistry Research*, vol. 60, no. 7, pp. 2971–2982, 2021.

**83** O.-T. Chis, J. R. Banga, and E. Balsa-Canto, "Structural identifiability of systems biology models: A critical comparison of methods," *PLoS One*, vol. 6, no. 11, p. e27755, 2011.

**84** H. Medina, N. Romano, and G. B. Chirico, "Kalman filters for assimilating near-surface observations into the Richards equation – part 3: Retrieving states and parameters from laboratory evaporation experiments," *Hydrology and Earth System Sciences*, vol. 18, no. 7, pp. 2543–2557, 2014.

**85** Y. Liu and H. V. Gupta, "Uncertainty in hydrologic modeling: Toward an integrated data assimilation framework," *Water Resources Research*, vol. 43, no. 7, 2007.

**86** R. F. Carsel and R. S. Parrish, "Developing joint probability distributions of soil water retention characteristics," *Water Resources Research*, vol. 24, no. 5, pp. 755–769, 1988.

**87** Y. Li, D. Chen, R. White, A. Zhu, and J. Zhang, "Estimating soil hydraulic properties of fengqiu county soils in the North China Plain using pedo-transfer functions," *Geoderma*, vol. 138, no. 3-4, pp. 261–271, 2007.

**88** M. Soet and J. Stricker, "Functional behaviour of pedotransfer functions in soil water flow simulation," *Hydrological Processes*, vol. 17, no. 8, pp. 1659–1670, 2003.

**89** A. Chatterjee, "An introduction to the proper orthogonal decomposition," *Current Science*, pp. 808–817, 2000.

**90** S. Debnath, S. R. Sahoo, B. T. Agyeman, and J. Liu, "Input-output selection for LSTM-based reduced-order state estimator design," *Mathematics*, vol. 11, no. 2, p. 400, 2023.

**91** S. Kung and D. Lin, "Optimal Hankel-norm model reductions: Multivariable systems," *IEEE Transactions on Automatic Control*, vol. 26, no. 4, pp. 832–852, 1981.

**92** A. C. Antoulas, *Approximation of large-scale dynamical systems*. SIAM, 2005.

**93** S. Gugercin and A. C. Antoulas, "A survey of model reduction by balanced truncation and some new results," *International Journal of Control*, vol. 77, no. 8, pp. 748–766, 2004.

**94** X. Cheng and J. Scherpen, "Gramian-based model reduction of directed networks," *arXiv preprint arXiv:1901.01285*, 2019.

**95** S. R. Sahoo, B. T. Agyeman, S. Debnath, and J. Liu, "Knowledge-based optimal irrigation scheduling of agro-hydrological systems," *Sustainability*, vol. 14, no. 3, p. 1304, 2022.

**96** S. R. Sahoo and J. Liu, "Adaptive model reduction and state estimation of agro-hydrological systems," *Computers and Electronics in Agriculture*, vol. 195, p. 106825, 2022.

**97** S. Debnath, B. T. Agyeman, S. R. Sahoo, X. Yin, and J. Liu, "Performance triggered adaptive model reduction for soil moisture estimation in precision irrigation," *International Journal of Adaptive Control and Signal Processing*, vol. 39, no. 8, pp. 1647–1662, 2024.

**98** M. Steinbach, G. Karypis, and V. Kumar, "A comparison of document clustering techniques," *CS&E Technical Reports*, 2000.

**99** A. P. Mazzini, E. N. Asada, and G. G. Lage, "Minimisation of active power losses and number of control adjustments in the optimal reactive dispatch problem," *IET Generation, Transmission and Distribution*, vol. 12, no. 12, pp. 2897–2904, 2018.

**100** E. J. de Oliveira, I. Da Silva, J. L. R. Pereira, and S. Carneiro, "Transmission system expansion planning using a sigmoid function to handle integer investment variables," *IEEE Transactions on Power Systems*, vol. 20, no. 3, pp. 1616–1621, 2005.

**101** B. T. Agyeman, S. R. Sahoo, J. Liu, and S. L. Shah, "An LSTM-based mixed-integer model predictive control for irrigation scheduling," *The Canadian Journal of Chemical Engineering*, vol. 101, no. 6, pp. 3362–3381, 2023.

**102** X. Ding and W. Du, "Optimizing irrigation efficiency using deep reinforcement learning in the field," *arXiv preprint arXiv:2304.01435*, 2023.

**103** W. Masson, P. Ranchod, and G. Konidaris, "Reinforcement learning with parameterized actions," *Proceedings of the AAAI Conference on Artificial Intelligence*, vol. 30, no. 1, 2016.

**104** B. T. Agyeman, M. Naouri, W. Appels, and J. Liu, "Irrigation management zone delineation and optimal irrigation scheduling for center pivot irrigation systems," *IFAC-PapersOnLine*, vol. 56, no. 2, pp. 9906–9911, 2023.

**105** B. T. Agyeman, M. Nouri, W. M. Appels, J. Liu, and S. L. Shah, "Learning-based multi-agent MPC for irrigation scheduling," *Control Engineering Practice*, vol. 147, p. 105908, 2024.

**106** R. L. Huffman, D. D. Fangmeier, W. J. Elliot, and S. R. Workman, "Irrigation principles," in *Soil and Water Conservation Engineering, Seventh Edition*, pp. 351–373, American Society of Agricultural and Biological Engineers, 2012.

**107** M. Schaarschmidt, A. Kuhnle, B. Ellis, K. Fricke, F. Gessert, and E. Yoneki, "Lift: Reinforcement learning in computer systems by learning from demonstrations," *arXiv preprint arXiv:1808.07903*, 2018.

**108** A. Richards and J. How, "Mixed-integer programming for control," in *Proceedings of the 2005, American Control Conference, 2005*, pp. 2676–2683, IEEE, 2005.

**109** B. T. Agyeman, J. Liu, and S. L. Shah, "Semi-centralized multi-agent RL for irrigation scheduling," *IFAC-PapersOnLine*, vol. 58, no. 14, pp. 145–150, 2024.

**110** W. J. Zhou, B. Subagdja, A.-H. Tan, and D. W.-S. Ong, "Hierarchical control of multi-agent reinforcement learning team in real-time strategy (RTS) games," *Expert Systems with Applications*, vol. 186, p. 115707, 2021.

**111** A. Tampuu, T. Matiisen, D. Kodelja, I. Kuzovkin, K. Korjus, J. Aru, J. Aru, and R. Vicente, "Multiagent cooperation and competition with deep reinforcement learning," *PLoS One*, vol. 12, no. 4, p. e0172395, 2017.

**112** C. S. De Witt, T. Gupta, D. Makoviichuk, V. Makoviychuk, P. H. Torr, M. Sun, and S. Whiteson, "Is independent learning all you need in the starcraft multi-agent challenge?," *arXiv preprint arXiv:2011.09533*, 2020.

**113** H. Nekoei, A. Badrinaaraayanan, A. Sinha, M. Amini, J. Rajendran, A. Mahajan, and S. Chandar, "Dealing with non-stationarity in decentralized cooperative multi-agent deep reinforcement learning via multi-timescale learning," in *Conference on Lifelong Learning Agents*, pp. 376–398, PMLR, 2023.

**114** D. Ernst, M. Glavic, F. Capitanescu, and L. Wehenkel, "Reinforcement learning versus model predictive control: A comparison on a power system problem," *IEEE Transactions on Systems, Man, and Cybernetics, Part B (Cybernetics)*, vol. 39, no. 2, pp. 517–529, 2008.

**115** R. R. Negenborn, B. De Schutter, M. A. Wiering, and H. Hellendoorn, "Learning-based model predictive control for Markov decision processes," *IFAC Proceedings Volumes*, vol. 38, no. 1, pp. 354–359, 2005.

**116** F. L. Da Silva and A. H. R. Costa, "A survey on transfer learning for multiagent reinforcement learning systems," *Journal of Artificial Intelligence Research*, vol. 64, pp. 645–703, 2019.

**117** L. P. Kaelbling, M. L. Littman, and A. R. Cassandra, "Planning and acting in partially observable stochastic domains," *Artificial Intelligence*, vol. 101, no. 1-2, pp. 99–134, 1998.

**118** R. A. Howard, *Dynamic Programming and Markov Processes*. Technology Press of Massachusetts Institute of Technology, Cambridge, MA, USA, 1960.

**119** M. B. Singer, D. T. Asfaw, R. Rosolem, M. O. Cuthbert, D. G. Miralles, D. MacLeod, E. A. Quichimbo, and K. Michaelides, "Hourly potential evapotranspiration at 0.1 resolution for the global land surface from 1981-present," *Scientific Data*, vol. 8, no. 1, p. 224, 2021.

**120** E. Léger, A. Saintenoy, Y. Coquet, P. Tucholka, and H. Zeyen, "Evaluating hydrodynamic parameters accounting for water retention hysteresis in a large sand column using surface GPR," *Journal of Applied Geophysics*, vol. 182, p. 104176, 2020.

**121** F. Abbasi, M. Javaux, M. Vanclooster, and J. Feyen, "Estimating hysteresis in the soil water retention curve from monolith experiments," *Geoderma*, vol. 189, pp. 480–490, 2012.

**122** K. Cao, S. H. Son, J. Moon, and J. S.-I. Kwon, "A closed-loop integration of scheduling and control for hydraulic fracturing using offset-free model predictive control," *Applied Energy*, vol. 302, p. 117487, 2021.

**123** D. R. Bennett and T. E. Harms, "Crop yield and water requirement relationships for major irrigated crops in southern Alberta," *Canadian Water Resources Journal*, vol. 36, no. 2, pp. 159–170, 2011.

**124** K. Saxton, W. Rawls, J. S. Romberger, and R. Papendick, "Estimating generalized soil-water characteristics from texture," *Soil Science Society of America Journal*, vol. 50, no. 4, pp. 1031–1036, 1986.

# Index




*Precision Irrigation for Agriculture: Integrating Machine Learning and Optimal Control Strategies*, First Edition. Bernard Twum Agyeman and Jinfeng Liu.

## r



## s



## t



## u



## v



## w



## z

Printed and bound by CPI Group (UK) Ltd, Croydon, CR0 4YY
17/09/2026
14964297-0007